RISK MANAGEMENT in Evaluating Mineral Deposits

By Jean-Michel Rendu

PUBLISHED BY THE
SOCIETY FOR MINING, METALLURGY & EXPLORATION

Society for Mining, Metallurgy & Exploration (SME)
12999 East Adam Aircraft Circle
Englewood, Colorado 80112
(303) 948-4200 / (800) 763-3132
www.smenet.org

The Society for Mining, Metallurgy & Exploration (SME) is a professional society whose more than 15,000 members represent professionals serving the minerals industry in more than 100 countries. SME members include engineers, geologists, metallurgists, educators, students, and researchers. SME advances the worldwide mining and underground construction community through information exchange and professional development.

Electronic edition published 2017.

ISBN: 978-0-87335-448-6
eBook: 978-0-87335-449-3

Library of Congress Cataloging-in-Publication Data

Names: Rendu, Jean-Michel, 1944- author.
Title: Risk management in evaluating mineral deposits / by Jean-Michel Rendu.
Description: Englewood, Colorado : Society for Mining, Metallurgy & Exploration, [2017] | Includes bibliographical references and index.
Identifiers: LCCN 2017012801 (print) | LCCN 2017017792 (ebook) | ISBN 9780873354493 | ISBN 9780873354486 (print)
Subjects: LCSH: Mine valuation. | Mineral industries--Risk management. | Mining engineering--Risk assessment.
Classification: LCC TN272 (ebook) | LCC TN272 .R463 2017 (print) | DDC 622.068/1--dc23
LC record available at https://lccn.loc.gov/2017012801

CONTENTS

PREFACE

Some of the largest investments and property acquisitions ever made by mining companies took place in the first decade of the twenty-first century. Many of these acquisitions were followed by write-downs of historical magnitude only a few years later. In 2013 alone, write-downs from six major international companies reached a total of $20 billion. These write-downs resulted in investors losing confidence in the mining industry; company shares losing significant value; and chief executive officers, top mining executives, and mining professionals losing their jobs.

These investments and subsequent write-downs are commonly attributed to the excitement resulting from the commodity super-cycle that characterized the 12-year period from 2004 to 2016. The first seven years depicted unprecedented commodity price increases, except for the short interruption following the 2008 global financial crisis. This bull market ended around 2011–2012 and was followed by a sharp decrease in prices for the next four years. One could point at the super-cycle to justify the risky decisions made in 2004–2010 and to explain the unprecedented write-downs that followed in 2011–2016. But one should not assume that prevailing economic conditions were the root cause of all flawed decisions. This would imply that management had no part in these determinations. It would mean that no lesson could be learned, that the mistakes made were unavoidable, and that the same mistakes will inevitably be made when the next economic cycle unavoidably occurs.

Why is it, for example, that a large base-metal mineral deposit was purchased for nearly $4 billion one year and written down less than three years later after updating the feasibility study? Why is it that a mining company paid a billion dollars to purchase mineral rights on a 1,000-km^2 exploration area and wrote down an even larger amount a few years later, after determining that the exploration potential was insignificant? Why is it that a mining project was approved for initial investment, including purchase of trucks, shovels, and major processing equipment, before having governmental agreement that mineral rights would be granted under reasonable terms and before evaluation of the mineral deposit had shown economic viability?

Was it because the market prices of minerals were increasing at a historically record rate at the time the investments were made, while falling just as fast or even faster a few years later when write-downs proved necessary? Was it because companies that did not take drastic actions to show growth and willingness to take risk were highly criticized by financial analysts and punished in the marketplace? Might it be that the pressure toward acquisition and development of new projects was such that overly optimistic outcomes were assumed and that the resulting increased risk of failure was either not recognized or largely ignored? Was it because the irrational exuberance that prevailed at the time encouraged managers to make rash decisions to show decisiveness?

Was it because industry-standard due diligence processes were bypassed? Is it possible that sellers were overvaluing the properties they offered and that buyers were overestimating their ability to create additional value? Was it because of management's overconfidence in the competence of the project team to accurately assess geological, mining, processing, infrastructure, legal, financial, environmental, political, and social risks? Was it because managers responsible for project evaluation were steered by the company's predetermined expectation of a positive outcome? Was it that a recommendation to go ahead with an acquisition would have resulted in immediate personal reward while the penalty for a project failure would have only been evident years later?

In the current environment, with lower and still depressed commodity prices, it may be difficult to understand how so many high-risk decisions could have been made only a few years ago. One must remember that prices were rushing up, a so-called super-cycle was in progress, demand for minerals was believed to be destined to increase continuously for the next decade or longer, and there was confidence that any unreasonable decision would be forgiven as a result of ever-increasing prices. Higher prices would make even the highest risk investments profitable.

It is a well-known but often ignored fact that price cycles have always characterized the mining business environment, and there is no reason to believe that this will change in the future. Stable prices are the exception. Even if at a given time all fundamentals are interpreted as predicting that long-term demand will continue to grow indefinitely, one should not forget that booms never last forever. Most mines require very high up-front investments that must be recovered over many years. Decisions must be made with full consideration of the fact that short- to medium-term price cycles will occur during the mine life, with both positive and negative consequences.

When considering the possible consequences of price cycles, one must keep in mind that costs of goods and services follow a similar but lagging pattern. During the early phase of the price super-cycle, investment decisions were often made under the assumption that the cost of goods and services would remain constant over time while the value of products sold would continue to increase. Would it not have been more reasonable to assume that not only shareholders, but also employees, suppliers, nearby communities, and governments would demand higher contributions from the economic benefits resulting from higher prices? The cost increases followed the rise in commodity prices, but the costs did not decrease proportionally when prices fell.

Why is it that decisions made by publicly traded mining companies are strongly influenced by the opinions of financial analysts, even though these analysts are not responsible for those decisions? Analysts do bring valuable insight, but are they more knowledgeable than the company management? Why is it that so many mining companies make similar decisions at the same time, such as risky investments during the upside of price cycles and disinvestment during the downside? Is it because there is perceived safety in following group decisions?

Mining companies lean toward following the same logic to make the same decisions at the same time. The result is an increase in peaks and troughs of price cycles. New mines are developed when prices increase. This is logical as cash flows, expected profits, and borrowing capabilities are improving. But bringing new mines into production takes years, and, more often than not, the result is overproduction when prices drop, thus increasing the magnitude and duration of this fall. When a mine is built and capital expenditures are sunk, production is maintained as long as cash flows remain positive.

Conversely lack of investment during the downside of the price cycle results in underproduction during the upside as well as sharper price increases. Leading managers are those who anticipate cycles and consistently make higher-valued decisions. But this implies making decisions that are counterintuitive at the time and losing the short-term comfort that comes from following group dynamics.

These are some of the issues that must be considered if the investment decision-making process is to be improved. And there is no better time to consider the preceding questions than during the tumultuous period the mining industry is currently going through, when consequences of some of the best and worse practices have been emphasized by arguably challenging worldwide economic conditions. Business cycles will continue to occur, and the timing of these cycles will continue to be unpredictable. We should attempt to learn from the past and not repeat the same mistakes. This book is intended to help move this learning process forward.

The objective is to provide guidelines that can be used to improve the decision-making process in a broad variety of circumstances. But each project is different, and it is appropriate to conclude with a quote from Alexis de Tocqueville in his 1848 introduction to *Democracy in America* (New York: Harper Perennial, 1988):

> I realize that despite the trouble taken, nothing will be easier than to criticize this book, if anyone thinks of doing so.
>
> Those who look closely into the whole work will, I think, find one pregnant thought which binds all its parts together. But the diversity of subjects treated is very great, and whoever chooses can easily cite an isolated fact to contradict the facts I have assembled, or an isolated opinion against my opinions.

I could not have said it better. But this does not absolve me from keeping full responsibility for errors, lapses, or absence of clarity that may be found in this book.

ACKNOWLEDGMENTS

I thank the Society for Mining, Metallurgy & Exploration who made publication of this book possible, including Jane Olivier, manager of book publishing, and Diane Serafin, managing technical editor.

I am grateful to Roussos Dimitrakopoulos, Francois Grobler, Steve Hoerger, Peter McCarthy, Harry Parker, and Pat Stephenson, who provided a number of helpful comments on an early draft of this manuscript.

And as always, I cannot thank my family enough for their love and support throughout my mining career.

CHAPTER 1

INTRODUCTION

The subject of mining is a very extensive one, and one very difficult to explain...This book indicates the length and breadth of the subject, and the number and importance of the sciences of which at least some little knowledge is necessary to miners.

Georgius Agricola, *De Re Metallica*, 1556[1]

This book is designed to inform decision makers of the complex processes that must guide mining companies in the evaluation and development of technically, financially, socially, and environmentally sound mining operations. *Decision-making* implies making a choice between options. Each option requires acquisition and analysis of information obtained from different sources. Project evaluation is a forward-looking process. Assumptions must be made concerning what the future will look like, at least partly, by looking at history. For each option, a model must be developed that represents how the project is expected to perform over time. The model must be not only deterministic, to estimate the project value under assumed conditions, but more appropriately probabilistic, to quantify the uncertainty with which this value is estimated. Choosing an option presents risks, which result

1 Quotes from Georgius Agricola's *De Re Metallica* are included in this book to show that the concerns of decisions makers in 1556 were similar to those we have today. Lou Henry Hoover, a linguist and Stanford University geology graduate, translated *De Re Metallica* from Latin to English in 1912. Her husband, Herbert Clark Hoover, a Stanford geology graduate, mining engineer, and U.S. president during the Depression years (1929–1932), contributed in placing the text in a modern mining context. There are interesting similarities between President Hoover's failure to get reelected in 1932, in the middle of the 1930s economic crisis, and the CEOs of mining companies losing their jobs during the 2010s commodity crisis.

from uncertainty concerning the information being used and the validity of the assumptions being made.

The model that represents the expected performance of a project is primarily a technical and financial model. The value of a project is first estimated taking into account technical feasibility, expected financial return, and risk. But what constitutes the "best" option may vary significantly depending on the perspective of the stakeholder who values the project. Shareholders might be primarily interested in the expected return on their investment. Governments might want to maximize employment and taxes. The main concern of local communities might be the environmental and social impacts.

A wide range of expertise is needed to evaluate a project, including geology, mining engineering, metallurgy, project engineering, social and environmental science, cost estimation, mineral economics, finance, taxes, legal and regulatory know-how, public and governmental relations, and project management. The value of expert contributions in determining a project's feasibility is highly dependent on efficient communications between disciplines. Such communications do not happen automatically. Specialists often do not have the time, opportunity, inclination, or necessary knowledge to visualize their role within the general context of a project evaluation. And communications are often compromised with each specialty being characterized by a language or terminologies of its own.

While specialists are expected to have extensive expertise in their own field, managers on the other hand are expected to have knowledge that crosses multiple disciplines. Managers must obtain reliable information from a variety of experts and interpret this information to make sound decisions. A manager must understand the link between all the parts that form the project, including technical, financial, environmental, and social. He or she must ensure the reliability and accuracy of the information received from all individuals working on the project. The value of choosing a specific option, and the risk that this choice entails, must be estimated without being influenced by preconceptions of what the correct answer should be.

The project manager must be able to communicate effectively with all those working on or responsible for the project, from specialists who contribute to the estimation of value and risk, to senior management and the board of directors (BOD) who must make final investment decisions. It is not unusual for the chief executive officer (CEO) of a mining company and members of the BOD to have little direct experience in the evaluation, construction, operation, and closure of mining projects. The CEO's responsibilities reside primarily in developing strategy and vision, setting the culture of the company, driving performance, building the team, allocating capital, approving objectives and business plans, determining what risk level is acceptable, and communicating with shareholders, financial institutions, governments, and the BOD. Unavoidably, information is lost during communications from specialists to the project manager, and ultimately to the CEO and the BOD. Processes and procedures should be established to formalize project evaluation and to avoid the loss of critical information, especially as it relates to the confidence that management should have in the estimation and control of the risk being taken.

Effective communication is necessary in two dimensions: horizontally between disciplines, and vertically between experts and successive levels of management. The need to ensure that this communication takes place should be recognized as a responsibility carried by all involved. Experts should avoid isolating themselves in the relative comfort of their specific expertise, where they are unlikely to be challenged. Managers should take time to understand the key disciplines whose input they rely on, to communicate with the experts in those disciplines, and to make sure that communications between disciplines are effective. Whenever possible, processes should be put in place to facilitate and monitor this information transfer. Ultimately, personalities play a critical role, and employing the right person at the right place, with the appropriate expertise and values, is a critical step toward ensuring a fair analysis of a project's feasibility.

There are instances when management appropriately controls or restricts internal or external communications to protect confidential information or to manage public disclosure of sensitive information. There are also unfortunate circumstances when information is filtered, with only that which supports personal or company objectives being communicated. Such situations can have technical, financial, legal, and ethical consequences of which management must be fully aware.

The remainder of this book is set out as follows. To evaluate a new project, and the risk associated with any decision made with respect to the project, it is helpful to first agree on general definitions of *risk*, *risk assessment*, and *risk management*, which is the subject of Chapter 2. It is also helpful to analyze situations where failure to appreciate the implication of decisions made had significant financial consequences. A number of examples are given in Chapter 3.

Project evaluation starts with exploration and, if successful, results in estimation of a mineral resource and a mineral reserve. Ultimately, a new mine might be developed. There are international guidelines that define the mineral assets, including resources and reserves, and how to publicly report such information. It is strongly recommended that mining companies follow these guidelines when internally evaluating projects. Publicly listed companies are required to follow country-specific regulations, which are, with few exceptions, based on these guidelines. The international guidelines are summarized in Chapter 4.

The value of a project, and the extent to which it is likely to contribute to a company reaching its objectives, is determined by modeling the expected project life cycle; estimating the expected financial, environmental, and social bottom lines; and quantifying the uncertainty associated with these estimates. Chapter 5 gives an overview of a typical life-of-mine cycle and the main risks likely to be faced when estimating this cycle.

Monte Carlo simulation and decision trees are methods commonly used to analyze uncertainty and guide the decision-making process in a risky environment. These tools are introduced in Chapters 6 and 7.

Risk or probabilistic variables can be classified according to whether uncertainty results from changes in space or over time. The geologic properties of a deposit are space related whereas future commodity prices are time related. This is discussed in Chapter 8.

The expected monetary value of a project is a critical component of what the project is worth to a mining company. But it is not the only component. Environmental and social impacts must also be taken into account. The risks associated with a project influence its value. Everything else being equal, two similar companies considering the same project will ascribe it different values depending on the extent to which they are risk averse, risk neutral, or risk loving. Interpretation of the same information will vary depending on the company that analyzes it. For these reasons, the term *utility* is used to distinguish between a project's expected financial value and the broader risk-adjusted value, which takes risk and opportunity into account. Chapter 9 shows how understanding the concepts of risk tolerance and utility function are essential to improve the decision-making process.

The sustainability of a mining project and its utility are best measured using full accounting and the triple bottom line: financial, environmental, and social. Analyzing a project using full accounting also represents an effective way to evaluate and mitigate risk. This is the subject of Chapter 10.

The preceding chapters form the foundation on which the rest of the book is built. The next chapter, Chapter 11, analyzes how the three bottom lines should be estimated in the context of a mining project.

In a mining company, the technical departments that supply the critical information needed to evaluate a project are geology, resource modeling, mining, processing, and infrastructure. Chapters 12 to 16 analyze the role played by these departments, the interaction needed between departments, the information they must develop, and how they contribute to the three bottom lines.

Chapter 17 provides an overview of the role of management in ensuring the success of a project evaluation exercise. A key role is to ensure the quality of the estimates and to control the risk associated with decisions concerning project development. Conditions that must be satisfied for management to be effective and the consequences of inadequate management practices are discussed in this chapter.

An overview of lessons learned is summarized in the last chapter, Chapter 18.

To illustrate the recommendations made in this book, examples are given of hypothetical situations where risky decisions were made that resulted in undesirable or unforeseen outcomes. The intent is to show the wide variety of pitfalls that may be encountered when analyzing projects, and how decisions made under specific conditions may have unpredictable consequences when conditions change.

Many examples were developed using publicly available information from several sources, including company news releases, quarterly and annual reports, reports from financial analysts, and publications from professional organizations. Other examples were specifically designed to reflect situations where the motivations of groups or individuals—governed by their own interests, values, and ethics—would conflict with the company's interests, values, and ethics.

These examples are not designed to analyze specific projects, or to comment on the rationale behind specific decisions made at a particular time, under precise conditions. None of the examples is meant to reflect on actions taken by specific mining companies or individuals at a specific time. Such actions may have been rational at the time they were taken. All examples contain a mix of real and potential elements, which have been combined to illustrate situations that are likely to present themselves in the future. Most decision makers will encounter similar situations during their professional lives. These examples are specifically designed to help decision makers be aware of such potentially risky situations and the need for adequate internal control structures and procedures for risk identification, prevention, and mitigation.

CHAPTER 2

MINING AS A COMPLEX AND RISKY BUSINESS

> *No person indeed can, without great and sustained effort and labour, store in his mind the knowledge of every portion of the metallic arts which are involved in operating mines.*
>
> Georgius Agricola, *De Re Metallica*, 1556

Mining projects are complex and risky, but so are projects in other industries. This is evident when one considers the well-advertised major bankruptcies that have plagued the automobile, information technology, imaging, real estate, banking, and other businesses over the years. Risk factors vary from one industry to another, but the approach to decision-making under uncertainty is fundamentally the same for all industries.

COMPLEXITY OF RESOURCE EVALUATION PROJECTS

Terry Williams, dean of Hull University Business School, defines a *complex project* as follows (ICCPM 2014):

> A project becomes particularly complex when it combines three effects: it is very complicated (lots of parts and lots of interconnections), it is highly uncertain (so there are likely to be many changes or disruptions) and it is heavily time-constrained (so there is no time to sit back and re-plan sensibly—disruptions need working-around immediately).

This description of complex projects well represents the situations encountered in the evaluation and development of mineral resources. Many variables of complex nature must be taken into account to evaluate a project. Some input variables, such as the grade of the mineralization, vary in space, while others, such as commodity prices, vary over time. Decisions must be made with incomplete information concerning these variables. Obtaining more information can reduce some of the project risk, but there is a cost in doing so. In many situations, decisions are time constrained and must be made quickly.

Even if one considers only technical components, such as geology, mining, and processing, there are very strong interconnections between these parts. Interconnections become even more complex when financial, managerial, environmental, and social aspects are added. The implication is that uncertainty concerning one part results in uncertainty pertaining to all other parts. The success of a mining project is dependent on the ability to assess the risk taken when making a decision, to predict what disruptions may occur during the life of the project and the likelihood of such occurrences, to have a plan of action to be taken if disruptions occur, and to determine whether the residual risk is acceptable.

RISK AND RISK ASSESSMENT

Risk is defined as the likelihood that the outcome of a project will differ from that which is expected. The three factors that must be taken into account to measure risk are

1. The probability that an event will occur,
2. The impact that this event will have on the project outcome, and
3. How the combination of probability and impact will be perceived by the various stakeholders.

The steps that must be followed to assess risk and make sound decisions are independent of the nature of the project and can be summarized as follows:

1. The decision-making company must have a clear understanding of its criteria for project evaluation. The perception of value is not the same for all stakeholders. Financial indicators, such as net present

value or payback period, are the primary—but not only—factors to be taken into account. Other indicators include those of a social and environmental nature. The same project will have a different value depending on the company's risk tolerance, a tolerance that may change over time. The term *utility* is used to define an extended definition of *value* which takes into account financial, social, and environmental indicators, as well as risk tolerance.

2. A determination must be made of the events or disruptions that could potentially influence, positively (upside) or negatively (downside), the utility of the project. Some of these events might have a high probability of occurrence, but the consequences may be of low significance. Events with a high probability of occurrence and significant consequences are those most likely to be taken into account when evaluating a project. It is easy to overlook critical events with low probability of occurrence but potentially major consequences (also called "black swan" events).

3. The probability that potential disruptions will occur must be quantified. These probabilities can be subjective, reflecting expert opinions, or objective, measured by statistical processes. Most probabilities are estimated using combinations of subjective and objective approaches.

4. After possible disruptions have been identified, one must determine what will be the impact on the project utility if they do occur. The cost/benefit of occurrence of a given event must be estimated taking into account the financial, environmental, and social consequences, which define the project utility.

These steps must be followed to assess risk and opportunities. Once completed, events with possible risky consequences should have been identified, together with the probability of occurrence of these events and the magnitude of their impact on the bottom line. Similarly, the probability of events with a positive impact on the project should have been characterized. Only then is it possible to consider risk-mitigating strategies, and the need for such mitigation and its cost. Such strategies should take into account not only ways to reduce negative impacts but also ways to improve positive impacts.

RISK MANAGEMENT

The International Organization for Standardization recommends the use of ISO 31000:2009, *Risk Management*, to help organizations increase the likelihood of achieving objectives, improve the identification of opportunities and threats, and effectively allocate and use resources for risk management. Options that should be considered include the following:

- **Avoiding the risk by deciding not to start or continue with the activity that gives rise to the risk.** A mining company might decide that uncertainty concerning an exploration project is too high to justify acquisition. Another company might decide that developing a project is not in its best interest because of high political risk.
- **Accepting or increasing the risk to pursue an opportunity.** Regional exploration is a high-risk activity to the extent that the probability of success is low, but the rewards can be very high. The risk of failure is accepted. While allocating more funds to exploration increases the cost of failure (more monies are at risk), such action is justified if there is a commensurate expected increase in probability of success.
- **Removing the risk source.** In some circumstances, fixed bids can be required to eliminate some financial risk sources. Long-term contracts or hedging can be used to reduce exposure to price and cost fluctuations. Eliminating risk related to market fluctuations reduces not only the negative but also the positive impact that these fluctuations may have on the project value.
- **Changing the likelihood.** Actions should be considered that could reduce the probability of occurrence of unfavorable events or increase the probability of favorable events. An inferred resource should not be the basis for a billion-dollar investment; it should first be drilled to the point that it can be converted to a measured or indicated resource. An infill drilling program may be an effective way to increase the likelihood that a mine will be able to feed the processing plant as planned.
- **Planning future mitigating actions.** If the risk is not imminent, are there actions that could be taken later that would mitigate the

risk? For long-life projects one may choose to postpone drilling of those parts of the deposit that will be mined later until such information is needed to finalize mine plans. If applicable, this mitigating strategy presents the advantage of delaying the cost of mitigation. These costs are incurred only if and when mitigation is deemed necessary. Postponing mitigation may be risky if such mitigation will be time-consuming. For example postponing drilling of an iron ore deposit may save current costs, but it would delay a rapid increase in production, which would be needed to benefit from a sudden increase in the price of iron ore.

- **Changing the consequences.** Could actions be taken now that would mitigate the impact of events that may occur in the future? For example, consider a mining project whose geometallurgical properties[1] are complex as well as technically and financially difficult to define. It might then be advisable to consider a processing plant that is expensive to build but flexible rather than a simpler plant that is only likely to perform reliably when receiving "typical" mineralization. As another example, if a project is located in an area where earthquakes are probable, construction methods should take this risk into consideration to minimize potentially disastrous consequences of an unlikely event.
- **Sharing the risk with another party or parties.** Contracts with employees, suppliers and customers, risk financing and partnering, and government participation are some methods of risk sharing.
- **Retaining the risk by informed decision.** Management may decide that the risk is acceptable, that it is not immediate and mitigating actions could be taken at a later date, or that the cost of risk mitigation is too high to be justified. When evaluating a mineral deposit, there is always a point where a decision is made to stop drilling and sampling; additional drilling will reduce geological uncertainty, a major risk factor in most mining projects, but cost and time requirements may be excessive.

1 *Geometallurgical properties* are geological characteristics of the deposit that are likely to influence, positively or negatively, the performance of a metallurgical process or the value of the product to be sold.

PUBLIC DISCLOSURE OF RISK IN MINING

The main difference between mining risk and the risk associated with projects in other businesses is the geologic component. Geology is what defines the location of mineral deposits, the properties of these deposits, whether they can be mined safely, and whether minerals of economic value can be feasibly extracted.

Discovering a deposit of potential economic value is the initial challenge faced by exploration companies. The probability of success is low, in most cases very low, but the rewards can be extremely high. Once a deposit has been discovered, considerable expenditures are progressively incurred to determine its economic value. If a project is not properly analyzed before development, the likelihood of failure may be significant and the cost of failure can be extremely high. To control risk, information must be acquired concerning the deposit and its environment. This information must be analyzed to define possible mining and processing methods, infrastructure requirements, and corresponding capital and operating costs. The potential social and environmental impact that a mining operation would have must be determined. The political, financial, and legal environment must be assessed. Obtaining the necessary information and processing it is expensive and time-consuming.

Some variables, such as the geologic properties of the deposit, vary in space, while others, such as the market conditions that will prevail when mining takes place, vary over time. Although investment of time and money can reduce the risk resulting from limited knowledge, there is a point where such investment is no longer justified and the remaining risk must be accepted. Predicting the future value of a time-related variable is particularly unreliable, and the effectiveness of the processes available to make such predictions is limited.

All publicly listed mining companies are required to include in their annual report a list of risk factors. This list varies among companies, but similarities are evident. The following is a list of typical risk factors included in such reports. Public release should include not only relevant factors but also the materiality of these factors as it relates to the various stakeholders.

- Financial risks:
  - Capital cost of developing the project (initial capital cost) and maintaining production to the end of life (sustaining capital)

  - Cost of operations (mining, processing, overhead, etc.), including change in cost of supplies and labor over time
  - Closure and rehabilitation costs, including differences between provisions and actual end-of-life costs
  - The price that will be received for the product sold (gold bullion, copper concentrate, washed coal, run-of-mine iron ore, etc.) at the time the sales take place
  - Change in terms concerning the availability and cost of supplies or the conditions under which the product can be sold
  - Inflation and changes in currency exchange rates
  - Ability to fund a capital-intensive project
- Technical and operational risks:
  - Reliability of the deposit's geological model
  - Reliability of the metallurgical assumptions made in developing the process
  - Operating performance of equipment, processes, and facilities (actual vs. planned)
  - Unexpected operational catastrophes
  - Failure of information technology systems
  - Failure or delays in supply lines
  - Access to and reliability of power supply, water, and consumables
- Environmental, social, and governmental risks:
  - Security of tenure to land, mineral rights, operating permits, and so on
  - Environmental impact and cost of mitigation
  - Safety, health, and community impacts

- Community relations and expectations, including expectation that costs and benefits of operations are fairly distributed
- Actions from nongovernmental organizations
- Labor relations
- Timing of approval for necessary governmental permits
- Changes in domestic and foreign laws and regulations
- Changes in tax laws and royalty agreements
- Natural catastrophes and climate change, which may impact operations and markets
- Domestic and international economic and political settings

MATHEMATICAL METHODS OF RISK ASSESSMENT

Mathematical methods for decision-making under uncertainty were developed five decades ago, if not earlier, and are still applicable today. Included are the following:

- Use of "decision trees" to graphically represent options and assist in deciding which action is most likely to result in reaching a desired objective
- Definition of the *utility* of a project as a measure to be used by decision makers to evaluate projects when risk tolerance and complex objectives are to be taken into account
- Use of a Monte Carlo simulation to combine uncertainty from a number of factors that influence the project utility
- Development of Bayesian statistics to compare risk as currently assessed with that expected to remain after more information is obtained

These methods are relatively simple and can be used by decision makers to effectively explain to management why specific recommendations are made. Recent advances in decision-making theory, and development of computerized systems to apply newly developed methodologies to ever more complex situations, have improved the ability to process very large data sets and compare multiple options. Whether one uses the simpler methods summarized earlier and further described in this book or more complex ones such as "real options analysis," the underlying principles are the same. The challenge in making investment decisions in mining does not reside in recognizing that mining is a risky business. The challenge is in properly evaluating this risk to decide how much risk is acceptable and to take proper action accordingly.

Decision trees, statistical modeling, Monte Carlo simulation, risk tolerance, and utility functions are discussed extensively in later chapters. But first it is appropriate to consider practical examples of situations where risky decisions are likely to be made with highly significant consequences and to determine what can be learned from such situations, the topic of Chapter 3.

CHAPTER 3

EXPENSIVE DECISIONS: WHAT MAY HAVE GONE WRONG?

> *There should be a certain limit of expenditures which miners should set themselves, lest blinded by the desire for excessive wealth, they throw all their money away.... Moreover, a prudent owner, before he buys shares, ought to go to the mine and carefully examine the nature of the vein, for it is very important that he be on his guard lest fraudulent sellers of shares should deceive him.*
>
> Georgius Agricola, *De Re Metallica*, 1556

One of the better ways to improve decision-making processes consists of learning from one's own failures and those of others. Questions were raised earlier concerning the rationale behind decisions made in the first decades of the twenty-first century, which resulted in project failures of unprecedented magnitude. Whether or not these decisions appeared rational at the time they were made is beside the point. What matters is why such decisions were made and how the decision-making process can be improved so that the same mistakes are not repeated over time.

In this chapter, several possible situations are discussed where decisions are likely to be made with very costly consequences. The objective is not to determine what might have gone wrong in specific situations that may have occurred in the past but rather to determine what factors are likely to contribute to inefficient decisions in similar situations that may be encountered in the future. Other examples are given in later chapters to add additional insight into specific aspects of the decision-making process.

These examples are based on information reflecting actual situations, whose facts were publicly disclosed by mining companies, financial analysts, or in technical publications. Many examples combine observations from several similar situations. All examples are intended to present realistic situations but are not intended to be factual. The intent here is not to describe or comment on specific projects or past decisions but to illustrate the importance of properly recognizing risk factors that can influence decisions in situations likely to be encountered.

EXAMPLE 1: UNDERESTIMATION OF POLITICAL RISK AND INFRASTRUCTURE COSTS

In this example, a company purchased a base-metal project in a remote part of Africa. Previous owners had extensively drilled the deposit. Due diligence was completed to evaluate the project's potential. It was recognized that the economic value of the project was dependent not so much on the geologic characteristics of the deposit, which were reasonably well understood, but rather on capital and operating costs, political uncertainty, and availability of a stable market on which to sell the concentrate that was to be produced. Some of the key risk factors included the following:

- Final agreement had not been reached concerning mineral rights, surface rights, tax rates, and profit sharing. Legal and political uncertainty implied considerable risk to project economic feasibility.
- Access to the mine site required construction of hundreds of kilometers of roads and railways across political boundaries. Legal and governmental negotiations were likely to delay project approval and construction start-up.
- There was no reliable geographic information concerning the areas on which construction would take place. The capital cost of road and railway lines was uncertain.

- Existing harbor facilities would also have to be upgraded. The country expected these facilities to benefit not only the mine but also other regional development. Cost sharing had to be negotiated.
- Whether and when agreement would be reached between the project owner and the countries involved was highly uncertain.

In spite of these uncertainties, the following decisions were made immediately after purchasing the property:

- Mining trucks, electric shovels, and critical milling equipment were ordered from manufacturers. The concern was that worldwide shortage of heavy equipment and long delivery times would otherwise result in unacceptable project delays.
- The hiring of key personnel that would be needed to operate the mine was started in earnest, the objective being to start development as soon as necessary permits were obtained, thus benefiting from prevailing high commodity prices.
- A detailed feasibility study was completed, based on preliminary estimates of the type of final agreements that would be reached between project owner and the many governmental entities whose approval was required.

Eventually the political issues were resolved. But at this point, the mining super-cycle had reached its inevitable end, metal prices had begun to fall, and the project was put on hold indefinitely. The cost of purchasing the property, buying equipment, completing the feasibility study, and hiring and eventually laying off critical personnel had to be written off. Within one year of the write-down, the company's chief operating officer, the project manager, and other senior officers of the purchasing company had been replaced. The following factors contributed to the project's failure:

- The buyer, the mining industry as a whole, investors, and financial analysts were overly optimistic that metal prices, which had significantly increased during the previous decade, would continue to do so in the foreseeable future.

- Insufficient consideration was given to the economic consequences of likely permitting delays and resulting increased probability of unfavorable economic conditions when production would start.
- The company was excessively optimistic concerning how quickly political and legal issues could be resolved, and what would be the cost of a final agreement.
- Due diligence was completed before purchasing the property. The feasibility study completed after acquisition showed that cost and time needed to develop the project had been significantly underestimated. Most significant was the error of estimation of the cost of infrastructure, including construction of roads and railways in difficult and poorly defined geographical, social, environmental, and political environments.
- Early purchase of equipment and hiring of key personnel was done to accelerate project development, thus increasing the expected net present value and the likelihood of benefiting from the ongoing high commodity prices. These decisions increased the cost of failure when political agreement was delayed and commodity prices failed to continue their upward trend.
- Early commitment to project development was made before reaching governmental agreement on permits, percentage ownership, tax rates, and royalties. The social license to operate had not been obtained from the local communities. This commitment decreased the bargaining power of the mining company while increasing that of government and communities, resulting in additional time delays and significant escalation of social and governmental costs.
- The cost of project failure had been significantly underestimated. This cost should have been measured not only by the magnitude of the possible write-down but also by the likely consequences of such a write-down, including lower share prices, shareholders and financial institutions losing confidence in management, and ensuing management changes.

- Without proper analysis, the probability of success was estimated as very high, and the magnitude of a successful outcome was significantly overestimated. Conversely, the likelihood of failure was estimated to be low and so was the cost of failure.

EXAMPLE 2: UNDERESTIMATION OF GEOLOGICAL AND GEOTECHNICAL RISK

In this example, a company purchased an underground mining operation that had been put on a care-and-maintenance program by the previous owner because of complex geology, unfavorable geotechnical properties, and resulting high operating costs and significant safety concerns. The new owner, who had considerable experience in underground operations under difficult geotechnical conditions, resumed operations shortly after purchase. However, operations continued to be technically and financially challenging, and the company had to close the mine within three years of acquisition. The project had to be written off in its entirety.

The following factors contributed to premature mine closure:

- Unfavorable geologic environment. The size, grade, and continuity of high-grade mineralized zones were much less favorable than expected at the time the due diligence was completed.
- Overestimation of the achievable production rate and underestimation of the operating costs. Because the continuity in high-grade mineralization was much lower than anticipated, smaller stopes had to be designed and more development was needed to access these stopes. This contributed to lower productivity and higher costs.
- Inadequate assessment of the geotechnical risk with significant consequences concerning operating costs and safety.
- Rock failures resulting in a number of underground incidents. Safety concerns were the main factors that resulted in the decision to close the mine.

Commodity prices and labor costs were relatively stable during the three years of mine operation and did not contribute to the decision to close down the mine. Unsafe mining conditions, low productivity, and high mining costs were the main contributors, which can all be attributed to geological and geotechnical factors. Reasons why the risk associated with these factors was not properly evaluated included the following:

- Limited amount of information made available by the seller to assess the value of the property. As often happens in such transactions, the seller's presentations gave a positively biased view of the deposit's potential. Even if business law gives some protection to the buyer, one should never forget the Latin warning: "Caveat emptor,"[1] let the buyer beware.
- Limited time available to analyze the available information. Acquisitions are typically time sensitive, a factor which benefits the seller and puts pressure on potential buyers to make quick and risky decisions.
- The buyer's wrong interpretation of the deposit's geology, with too much reliance given to the seller's interpretation. While earlier operations had shown a very complex spatial distribution of high-grade mineralization, the new interpretation assumed high continuity between widely spaced drill holes. Experienced modelers should know that if geologic complexities are ignored, widely spaced drill-hole intercepts are easily used to produce highly continuous but unrealistic models of mineralization.
- Wrong interpretation by the buyer of the geotechnical environment and its impact on safety, mining rate, and operating costs. Lack of information contributed to an optimistic interpretation, which should have been tempered by the well-known challenges that had led the previous owner to close the mine.
- Overconfidence on the part of the buyer in the strength of its experience and capabilities to mitigate any difficulty encountered once restarting the mine. This reflects a rather common, but not always

1 The principle of caveat emptor can be traced back to 500 BC Roman law. The first unanimous U.S. Supreme Court decision that laid down the rule of caveat emptor in U.S. law was written on March 15, 1817, in *Laidlaw v. Organ* (15 U.S. 178).

justified, belief on the part of buyers that they can add value to a project beyond that recognized and presented by the seller.

EXAMPLE 3: OVERESTIMATION OF EXPLORATION POTENTIAL

This example analyzes the factors that may have influenced decisions made in a historical exploration project as well as the consequences of these decisions. The project description is based on publicly available information; however, no inference should be made concerning the extent to which these factors actually influenced decisions made in the past.

A mining company spent $1 billion to acquire mineral rights in a 1,000-km^2 tract of land in a remote area with difficult access and lacking infrastructure. Factors that influenced this acquisition included the following:

- Geological environment. The buyer viewed the environment as favorable to the discovery of a number of large deposits whose development would justify the cost of acquisition.
- Size of exploration area. Statistically speaking, given a favorable geologic environment, the probability of discovering significant mineralization increases with the size of the area under consideration. While under-explored, a significant part of the acquired area was considered as having high geologic potential.
- Presence of known mineral resources, which had been drilled by previous owners but not fully evaluated. The geologists were confident that the size of these resources could be expanded.
- Confidence in the financial and technical ability to add value. The acquiring company was also confident that, given its financial resources and the strength of its exploration team, it would be able to discover new mineral deposits, which the seller had failed to do.

Some financial analysts considered the cost of this acquisition as extreme and criticized the decision accordingly. Following the acquisition, the company completed a pre-feasibility study aimed at further defining and developing the known mineralization. Exploration of surrounding areas was also intensified. The results obtained can be summarized as follows:

- The potential for significant additions to the known and partially drilled mineral resources did not materialize.
- The capital cost of developing the area, which would require construction of railroads, shipping facilities, high-maintenance roadways, and a local airport, had been underestimated.
- Updated estimates of the operating costs were significantly higher than predicted at the time of acquisition.
- The acquisition was based on the expectation that large deposits would be discovered. This assumption was disproved when a better understanding of the geology was acquired.
- With the discovery of significant mineralization eventually considered unlikely, detailed exploration of the 1,000-km^2 property was no longer considered economically justified.

As a result of these changes in geologic understanding and better estimation of capital and operating costs combined with falling metal prices, the company had to write off the entire cost of acquisition, plus exploration and development costs that were incurred following this acquisition. Factors likely to have contributed to project failure can be summarized as follows:

- Industry-wide optimism concerning future metal prices that prevailed at the time of acquisition (mining companies were encouraged to make risky decisions to increase their production profile)
- Overestimation of the economic potential of the deposits that had been discovered by the previous owners
- Overestimation of the geologic potential, defined as the probability that large deposits were present within the exploration area
- Underestimation of the cost of exploring the area and overestimation of the likelihood that significant mineralization could be discovered at a reasonable cost, even if such mineralization was present

- Underestimation of the capital and operating cost of developing new mines in a particularly difficult geographic environment
- Underestimation of the cost of development and exploration failure, measured not only by the magnitude of any write-down but also by the likely consequences of such a write-down, including lower share prices, shareholders and financial institutions losing confidence in management's ability to make sound investment decisions, and ensuing management changes

EXAMPLE 4: FAILED COMMUNICATION BETWEEN GEOLOGIST AND METALLURGIST

This last example relates to a project whose capital costs significantly exceeded those estimated during the pre-feasibility study, because of unexpected geometallurgical and geochemical complexities that resulted in costly plant redesign and production delays. The deposit had been extensively drilled, and detailed geological, geochemical, mineralogical, and geotechnical studies had been completed. Metallurgical tests were completed on core samples, which showed that whole-ore roasting would be the preferred processing method. A preliminary plant design was developed accordingly. To be able to finalize the design, the process engineers determined that a significant tonnage of high-grade material would have to be tested in a pilot plant. A request for the necessary material was made to the project geologist.

The requested tonnage of high-grade material could only be obtained from an active underground exploration drift. The material was tested in the pilot plant, and a detailed design of the most suitable roasting facility was developed accordingly. This was the last task that had to be completed before the project feasibility study could be finished.

A final meeting of the project team was planned over a two-day period during which geologists, mining engineers, metallurgists, and process engineers were to review all technical aspects. The following details quickly became apparent:

- The material selected for metallurgical testing represented "high-grade" material as understood by the geologist. A preliminary mine plan showed there was not enough material of such grade to justify a project.

- For sufficient tonnage to be delivered to the roaster, a lower cutoff grade would be required. The roaster head grade would be significantly lower than that of the material tested in the pilot plant.
- The term *high grade* did not mean the same thing to the geologists ("material significantly higher grade than average"), the mining engineers ("material of sufficient quantity and quality, which the mine could deliver consistently to the processing plant"), and the process engineers ("material representative of expected plant feed that could be processed efficiently").
- The material to be roasted was geochemically complex, and the plant design included facilities to prevent release to the environment of deleterious elements. These facilities were designed on the basis of the percentage of such elements in the material that was tested. During the review meeting, the geologists pointed out that this high-grade material contained much lower percentages of deleterious elements than the lower-grade material that was to represent the bulk of the plant feed. The designed emission control facilities would have to be redesigned to satisfy environmental regulations.
- The pilot plant tests and roaster plant design had been completed over an eight-month period. Obtaining new representative lower-grade material, testing it, and modifying the design would result in an additional ten-month delay and considerable extra costs.
- The geochemical properties of lower-grade material were significantly more complex than those of the high-grade material that had been tested. The estimated capital and operating costs of the redesigned processing and environmental control facilities would be significantly higher than those that had been estimated on the basis of the first series of tests.

LESSONS LEARNED

These examples show the wide variety of factors that can contribute to failure in properly estimating the feasibility of mineral projects. Some of these factors, such as commodity prices, are changing over time and cannot be controlled by the project owner. However, the probability of significant changes in prices can be estimated and simulated, which helps better inform the decision process. Other factors, such as the buyer's understanding of the geology of the deposit, might be, to some extent, controllable. Additional drilling can be used to improve quantity and quality of geologic information. Relying on experts can result in a more reliable interpretation of available data. There is a cost to reducing uncertainty, and analyses must be completed to determine whether the expected reduction in project risk is sufficient to justify this cost. In acquisition and project development, time is of the essence, and decisions are made with limited information, which can considerably increase risk. Additional risk factors are of a human nature: overconfidence in one's expertise as compared to that of others; poor communications between experts; and not properly assessing which factors may influence the likelihood of success or failure.

Before continuing this analysis of the risks associated with mining projects, and to avoid confusion that can result from misunderstanding of the terminology being used, one must acquire clarity with respect to what is meant by *exploration results*, *mineral resources*, and *mineral reserves*. Fortunately, there exists global agreement on this terminology, which is summarized in the next chapter.

CHAPTER 4

DEFINITION AND PUBLIC REPORTING OF MINERAL ASSETS

> *It is complained that some sellers and buyers of the shares in mines are fraudulent...Indeed, a wise and prudent man, skilled in this art, if he doubts the trustworthiness of a seller or buyer, goes at once to the mine that he may for himself examine the vein that has been so greatly praised or disparaged, and may consider whether he will buy the shares or not. But people say, though such an one can be on his guard against fraud, yet a simple man and one who is easily credulous, is deceived.*
>
> Georgius Agricola, *De Re Metallica*, 1556

Effective communication requires a common language and trust in the information being communicated. This is true between individuals in their daily lives as well as between mining professionals when evaluating projects. It is also true within and between mining companies and their shareholders and other stakeholders. A number of national and international guidelines and regulations have been developed to improve communication related to mineral assets. These guidelines and regulations define terminology that must be used, and propose or specify studies that must be completed before a certain terminology can be applied, as well as the format that must be followed when publicly reporting information.

THE CRIRSCO INTERNATIONAL REPORTING TEMPLATE

Mining professionals and the institutes and societies that represent them have long recognized the need to standardize public reporting of information concerning exploration results, mineral resources, and mineral reserves. This led to the formation of an international committee known since 2001 as CRIRSCO, the Committee for Mineral Reserves International Reporting Standards. CRIRSCO's role is to promote high standards to be used by the mining industry when reporting mineral deposit estimates (mineral resources and mineral reserves[1]) and exploration progress (exploration results). Formed in 1994, this committee is recognized worldwide—by regulators, professional societies, exploration companies, mining companies, the United Nations, and consulting and accounting firms—as the key international organization representing the mining industry on issues relating to the classification and reporting of mineral assets.

CRIRSCO is composed of representatives of organizations, known as national reporting organizations (NROs), which are responsible for developing mineral reporting codes and guidelines. At the time of this writing, CRIRSCO members included representatives from Australasia, Brazil, Canada, Chile, the European Union, Kazakhstan, Mongolia, Russia, Southern Africa, and the United States. A number of other countries are considering membership. One of CRIRSCO's key contributions is the publication and maintenance of the *CRIRSCO International Reporting Template for the Reporting of Exploration Results, Mineral Resources and Mineral Reserves* (CRIRSCO 2013). NROs wishing to become members of CRIRSCO are required to use this template when developing national reporting codes or guidelines. Because of differences in the legal and regulatory environment of each country, national reporting codes or guidelines have been developed that follow the CRIRSCO template but contain additional information, requirements, and clarifications[2]. All companies and individuals responsible for the estimation and public reporting of mineral assets should make themselves fully conversant of the content of the international CRIRSCO template, the

1 Although terminology differs between countries because of language and/or traditions, definitions, and meaning remain the same.

2 National reporting codes or guidelines can be accessed through the CRIRSCO website.

applicable national codes and guidelines, and other rules and regulations specific to the country where reporting is to take place.

In this chapter, a summary is given of the international definitions of *exploration results*, *mineral resources*, and *mineral reserves*, and the studies that must be completed before reporting such information. The required competence of the person or persons responsible for estimation and reporting of mineral assets is discussed. Significant portions of this chapter are direct quotes from the 2013 *CRIRSCO International Reporting Template*. This summary is designed to introduce the reader to internationally accepted definitions and concepts that are used throughout this book. Considerable additional information can be found in the *CRIRSCO International Reporting Template* and the national codes and guidelines. Many of the terms in this chapter including, but not limited to, *exploration results*, *mineral resources*, and *mineral reserves*, are specifically defined in the CRIRSCO template and other national codes, regulations, and guidelines.

COMPETENCE AND RESPONSIBILITY

The main principles governing design and application of the CRIRSCO template, and derivative reporting codes, standards, and guidelines (collectively known as CRIRSCO-type reporting standards) are "transparency, materiality and competence." To ensure that these principles are followed, and as required by most national stock exchange regulators, the persons estimating and reporting mineral resources and mineral reserves must satisfy specific competence requirements. A *competent person*[3] is defined as a minerals industry professional belonging to a recognized professional organization that has enforceable disciplinary processes including the powers to suspend or expel a member. A competent person must have a minimum of five years[4] of relevant experience in the style of mineralization or type of deposit under consideration and in the activity undertaken by that person.

3 The term *competent person* (CP) is used in Australia, the United States, Europe, and South Africa. Similar terms are used in other countries, such as *qualified person* (QP) in Canada, *persona competente calificada* in Chile, and *profissionais qualificados* in Brazil.

4 At the time of this writing, consideration was given to increasing this minimum number to seven years.

If the competent person is preparing a report on exploration results, the relevant experience must be in exploration. If the competent person is estimating or supervising the estimation of mineral resources, the relevant experience must be in the estimation, assessment, and evaluation of mineral resources. If the competent person is estimating or supervising the estimation of mineral reserves, the relevant experience must be in the estimation, assessment, evaluation, and economic extraction of mineral reserves.

The requirement of five years' relevant experience is a minimum and additional experience may be required depending on the scope of work and possible consequences of recommendations to be made by the competent person. For example, while five years is likely to be sufficient to add mineral reserves to an existing operation, a mining company would be wise to expect significantly more than five years' relevant experience if the project under consideration is a new mining operation with capital cost expected to be in the billion dollars range.

The term *competent person* or its equivalent is accepted worldwide to describe the individual responsible for a part or the whole of the estimation and documentation of exploration results, mineral resources, and mineral reserves, which form the basis for public reports. The term *subject matter expert* is often used to describe an individual who exhibits the highest level of expertise in performing a specialized job, task, or skill within an organization. The two terms are not interchangeable. A subject matter expert does not necessarily satisfy all the conditions to be recognized as a competent person. Only a competent person can take responsibility for the information that is the basis for the public release of information related to mineral assets.

DEFINITIONS OF EXPLORATION RESULTS, MINERAL RESOURCES, AND MINERAL RESERVES

Figure 4.1 shows the relationship between exploration results, mineral resources, and mineral reserves. Information needed before exploration results, mineral resources, and mineral reserves are reported increases from top to bottom and from left to right. Mineral resources must be classified in function of increasing confidence, from *inferred* to *indicated* and *measured.* Only indicated and measured resources can be converted directly to a mineral

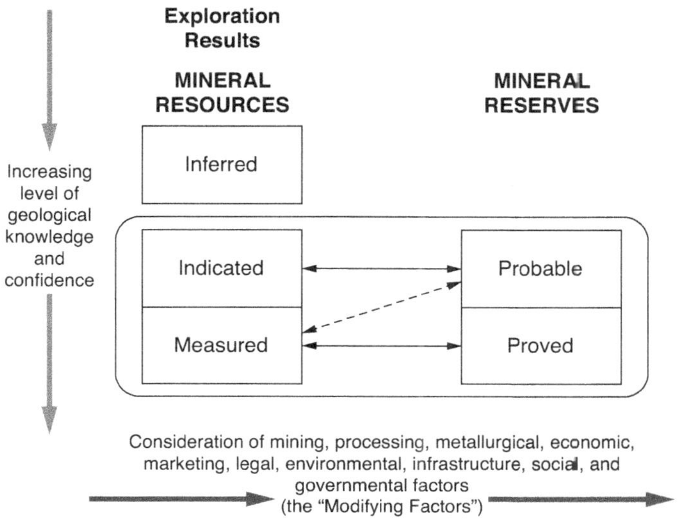

Source: CRIRSCO 2013

FIGURE 4.1 General relationship between exploration results, mineral resources, and mineral reserves

reserve. Depending on the level of confidence, mineral reserves must be classified as *probable* or *proved*. It is not always possible to convert a mineral resource to a mineral reserve.

Exploration results should be of sufficient interest that they might eventually lead to the estimation of mineral resources. To estimate a mineral resource, one must have sufficient geological information of high enough quality to be able to develop a deposit model and to estimate tonnages and grades that show reasonable prospects for eventual economic extraction. The need to show reasonable prospects for eventual economic extraction implies that preliminary estimates of likely mining and processing methods and costs have been considered, as well as other matters such as infrastructure requirements and likely value of product sold. In other words, an increased level of geological knowledge and confidence is necessary before one can declare a mineral resource, but it is not sufficient by itself. Reasonable technical and financial assumptions must also be made. To convert a mineral resource to a mineral reserve, one must consider modifying factors whose knowledge and understanding is sufficient to be able to state with a high enough level of confidence that economic development of the project is justified.

Exploration Results

Exploration results include data and information generated by mineral exploration programs that might be of use to investors but which do not form part of a declaration of mineral resources or mineral reserves. During the early stages of exploration, the quantity or quality of data available is generally not sufficient to allow reasonable estimates of tonnage and grade, and any such estimates undertaken for internal purposes must not be publicly reported. However, descriptions of exploration targets or exploration potential can be publicly discussed if expressed so as not to misrepresent them as an estimate of mineral resources or mineral reserves.

Mineral Resources

A *mineral resource* is a concentration or occurrence of solid material of economic interest in or on the earth's crust in such form, grade, or quality and quantity that there are reasonable prospects for eventual economic extraction. The location, quantity, grade or quality, continuity, and other geological characteristics of a mineral resource are known, estimated, or interpreted from specific geological evidence and knowledge, including sampling. Mineral resources are subdivided, in order of increasing geological confidence, into inferred, indicated, and measured categories. Portions of a mineral deposit that do not have reasonable prospects for eventual economic extraction must not be included in a mineral resource. Material assumptions made in determining the reasonable prospects for eventual economic extraction should be clearly stated in public reports of mineral resources.[5]

5 While not a CRIRSCO requirement, the prospects for eventual economic extraction are best assessed by completion of a study that is at least at the level of a scoping study, as defined later in this chapter.

Mineral Reserves

A *mineral reserve* is the economically minable part of a measured and/or indicated mineral resource. It includes diluting materials and allowances for losses, which may occur when the material is mined or extracted. It is defined by studies at the pre-feasibility or feasibility level, as appropriate, that include application of modifying factors. Such studies demonstrate that, at the time of reporting, extraction could reasonably be justified. The reference point at which mineral reserves are defined, usually the point where the ore is delivered to the processing plant, must be stated. Mineral reserves are subdivided, in order of increasing confidence in the modifying factors, into probable and proved categories.

Modifying factors are considerations used to convert mineral resources to mineral reserves. These include, but are not restricted to, mining, processing, metallurgical, infrastructure, economic, marketing, legal, environmental, social, and governmental factors. Confidence in the modifying factors determines whether a mineral reserve can be declared and whether it can be classified as probable or proved.

Inferred resources cannot be converted to mineral reserves because the uncertainty of the geological framework is too high to support mine planning, including mining method and estimates of dilution, and ore loss. Inadequate metallurgical test work may prevent formulation of realistic processing alternatives. The quality of available information is not sufficient to support a comprehensive study of a range of options for the technical and economic viability of the mineral project.

CLASSIFICATION OF MINERAL RESOURCES AND MINERAL RESERVES

Mineral resources and mineral reserves must be classified in order of increasing confidence. The classes of mineral resources are inferred, indicated, or measured, and those of mineral reserves are probable or proved. These classifications are designed to communicate the level of confidence that one should place in the estimates. While there is an element of subjectivity in the classification of mineral resources and mineral reserves, the role of the competent person in making these classifications should not be taken lightly, as regulators require them to be published, and only measured and indicated resources can be converted to mineral reserves.

Classes of Mineral Resources

An *inferred mineral resource* is that part of a mineral resource for which quantity and grade or quality are estimated on the basis of limited geological evidence and sampling. Geological evidence is sufficient to imply, but not verify, geological and grade or quality continuity. An inferred mineral resource has a lower level of confidence than that applying to an indicated mineral resource and must not be converted directly to a mineral reserve. It is reasonably expected that the majority of inferred mineral resources could be upgraded to indicated mineral resources with continued exploration.

An *indicated mineral resource* is that part of a mineral resource for which quantity, grade or quality, densities, shape, and physical characteristics are estimated with sufficient confidence to allow the application of modifying factors in sufficient detail to support mine planning and evaluation of the economic viability of the deposit.

Geological evidence is derived from adequately detailed and reliable exploration, sampling, and testing and is sufficient to assume geological and grade or quality continuity between points of observation. An indicated mineral resource has a lower level of confidence than that applying to a measured mineral resource and may only be converted to a probable mineral reserve.

A *measured mineral resource* is that part of a mineral resource for which quantity, grade or quality, densities, shape, and physical characteristics are estimated with confidence sufficient to allow the application of modifying factors to support detailed mine planning and final evaluation of the economic viability of the deposit.

Geological evidence is derived from detailed and reliable exploration, sampling, and testing and is sufficient to confirm geological and grade or quality continuity between points of observation. A measured mineral resource has a higher level of confidence than that applying to either an indicated mineral resource or an inferred mineral resource. It may be converted to a proved mineral reserve or a probable mineral reserve.

Classes of Mineral Reserves

A *probable mineral reserve* is the economically minable part of an indicated and, in some circumstances, a measured mineral resource. The confidence in the modifying factors applying to a probable mineral reserve is lower than that applying to a proved mineral reserve. A probable mineral reserve has a lower level of confidence than a proved mineral reserve but is of sufficient quality to serve as the basis for a decision on the development of the deposit.

A *proved mineral reserve* is the economically minable part of a measured mineral resource. A proved mineral reserve implies a high degree of confidence in the modifying factors. A proved mineral reserve represents the highest confidence category of reserve estimate.

DEFINITION OF SCOPING, PRE-FEASIBILITY, AND FEASIBILITY STUDIES

Before a mineral resource is declared, the mining company should[6] complete a study that is at least at the level of a Scoping Study, which CRIRSCO defines as follows:

> A Scoping Study is an order of magnitude technical and economic study of the potential viability of Mineral Resources that includes appropriate assessments of realistically assumed Modifying Factors together with any other relevant operational factors that are necessary to demonstrate at the time of reporting that progress to a Pre-Feasibility Study can be reasonably justified.

A mineral reserve can only be declared after a study that is at least at the level of a pre-feasibility or a feasibility study has been completed, which

6 CRIRSCO does not require that a scoping study be completed before declaring a mineral resource. Such a study should however be considered best practice to adequately document mineral resources that are to be publicly reported.

demonstrates that a mine can be developed economically. A *pre-feasibility study* is defined as follows:

> A Pre-Feasibility Study is a comprehensive study of a range of options for the technical and economic viability of a mineral project that has advanced to a stage where a preferred mining method, in the case of underground mining, or the pit configuration, in the case of an open-pit, is established and an effective method of mineral processing is determined. It includes a financial analysis based on reasonable assumptions on the Modifying Factors and the evaluation of any other relevant factors which are sufficient for a Competent Person, acting reasonably, to determine if all or part of the Mineral Resource may be converted to a Mineral Reserve at the time of reporting. A Pre-Feasibility Study is at a lower confidence level than a Feasibility Study.

A *feasibility study* is defined as follows:

> A Feasibility Study is a comprehensive technical and economic study of the selected development option for a mineral project that includes appropriately detailed assessments of applicable Modifying Factors together with any other relevant operational factors and detailed financial analysis that are necessary to demonstrate at the time of reporting that extraction is reasonably justified (economically mineable). The results of the study may reasonably serve as the basis for a final decision by a proponent or financial institution to proceed with, or finance, the development of the project. The confidence level of the study will be higher than that of a Pre-Feasibility Study.

Depending on regulations that are country specific, either a pre-feasibility or a feasibility study may be required before a mineral reserve is declared. Even if a feasibility study is not required by regulations, the decision whether a feasibility study should be completed remains with the competent person responsible for the estimate of mineral reserves. A high-risk project or one whose economic feasibility is marginal is likely to require completion of a feasibility study.

PUBLIC REPORTING OF RISK FACTORS

When publishing exploration results, mineral resources, and mineral reserves, additional information may have to be included concerning the level of confidence the reader can have in these estimates. This information may relate to upward potential or downward risk. Expectation of public release of such information is reflected by the CRIRSCO governing principles of transparency, materiality, and competence. International and national accounting rules, including the U.S. generally accepted accounting principles (GAAP), require publication of relevant information and inclusion of a "Risk Factors" section in company annual reports.

Risk factors must be taken into account before deciding if estimates of mineral assets can be publicly released, and whether these estimates are of sufficient quality to be used as a basis for capital investment. These risks, which are discussed in detail in subsequent chapters, include the following:

- Exploration results are subject to errors, and the geologic information may not be sufficient to properly assess the value of a mineral deposit.
- Interpretation of available geologic information may not be correct and is likely to change as more information becomes available.
- Deposit models, which are used to develop mine plans, production schedules, and cash flow estimates, are only approximations of reality.
- Classification of mineral resources as measured, indicated, or inferred contains a significant amount of subjectivity.
- Overconfidence in the deposit model may result in a mineral resource being classified as measured or indicated when it should be considered as inferred. Such an error would result in an overestimation of reserves and an underestimation of the risk associated with these reserves.
- A mineral resource can only be reported as a resource if appropriate conceptual analyses have been completed to show reasonable prospects for eventual economic extraction.

- Modifying factors that must be considered before declaring a mineral reserve include mining, processing, metallurgical, infrastructure, economic, marketing, legal, environmental, social, and governmental factors. These factors are estimated at the time the project is evaluated. Such estimates are subject to errors resulting from lack of reliable information, incorrect interpretation of this information, or unexpected changes that may occur over time.
- The value of key factors, such as costs and value of product sold, is likely to change over the life of the project. Future values are likely to differ—sometimes significantly—from current estimates.
- The results of the pre-feasibility or feasibility study that justifies declaration of a reserve are probabilistic in nature. The risk that the project will not perform as expected must be assessed and clearly communicated to management, shareholders, and other stakeholders.
- Statements concerning mineral resources and mineral reserves are forward looking. These statements are based on prediction of future events that may not materialize as predicted.

MANAGEMENT'S RESPONSIBILITIES

Management gives considerable attention to the estimation and public reporting of exploration results, mineral resources, and mineral reserves. Management must decide whether, and if so, when such estimates can or must be reported. Public release, or lack of it, can have significant legal as well as financial repercussions. The value of a mineral asset is highly dependent on whether it is classified as a measured, indicated, or inferred resource, or a proved or probable reserve. Only measured and indicated resources can be taken into account when a pre-feasibility or feasibility study is used to justify public release of proved and probable reserves. While, for internal purposes, a mining company may specify how much measured, indicated, and inferred material can be included in a mine plan before a decision is made to develop a new mine or expand an existing one, inclusion of inferred resources is not permitted for public reporting of mineral reserves.

One of the key responsibilities of management is to ensure that project studies are completed with full integrity, taking into account regulatory requirements and company values. Effective internal control structures and procedures must be established and maintained to ensure that all codes and regulations are followed, to prevent the use of unacceptable practices, and to ensure that if such practices are taking place, they are effectively detected and corrective action is immediately taken.

Those who are responsible for discovering a mineral deposit, analyzing it, and bringing it to production are often rewarded for their achievement. If properly designed, rewards such as those included in bonus plans are strong motivations to complete studies efficiently and in a timely fashion. Conversely, rewards can motivate individuals to underestimate the risk associated with a project and recommend unwise investment decisions. Management should not give the impression that there is a "right" and a "wrong" answer to a study. Failure to reach the "right" answer should not result in management rejecting the answer as unacceptable. The "wrong" answer may be the correct answer.

According to Section 404 of the Sarbanes–Oxley Act of 2002, the Securities and Exchange Commission requires that each annual report contain an internal control report, which

1. States the responsibility of management for establishing and maintaining an adequate internal control structure and procedures for financial reporting; and
2. Contains an assessment, as of the end of the most recent fiscal year of the issuer, of the effectiveness of the internal control structure and procedures of the issuer for financial reporting.

Mineral assets play a key role in the valuation of mining companies. Section 404 is understood to apply to the public reporting of exploration results, mineral resources, and mineral reserves. Mineral assets and exploration potential are used in the application of GAAP for determination of value beyond proved and probable reserves, or VBPP, as it applies to purchase price allocation and impairment testing (FASB 2008; Graff 2008).

CHAPTER 5

LIFE-OF-MINE CYCLE AND RISK FACTORS

> *Those who think that mining is of no advantage to the men who follow the occupation assert...that scarcely one in a hundred who dig metals or other such things derive profit therefrom.... But persons who hold these views do not perceive how much a learned and experienced miner differs from one ignorant and unskilled in the art.*
>
> Georgius Agricola, *De Re Metallica*, 1556

In earlier chapters, mining as a risky business was discussed. The need for appropriate competence to evaluate mineral assets was emphasized, and the three stages that must be successfully completed before a project is considered suitable for economic development were defined—exploration, estimation of mineral resources, and estimation of mineral reserves. Now attention must be brought to a typical mine life cycle. It is this mine life that must be predicted to determine whether a project is viable. It is in the context of this mine life that the project risk can be determined.

DEFINITION OF RISK, RISK FACTORS, AND RISK MANAGEMENT

In Chapter 2 it was argued that evaluation of mineral assets could be classified as a "complex problem" for the following reasons:

- Mineral projects have many parts that are interconnected and whose analysis requires a wide variety of expertise.

- There is a high level of uncertainty with respect to the outcome of a project evaluation. Information concerning the current and future value of critical input variables is limited.
- Conditions will change during the life of the project that are not predictable, and rapid decisions will have to be made to adapt the project to these changes.

Risk is broadly defined as the probability that an event will occur, and the consequences that this event would have if it occurred. Consequences can range from positive to negative. Common practice consists of defining *risk* as reflecting the probability of negative consequences while *opportunities* refers to positive consequences. Project risk is a function of the range of values that the project might take, and the probability that these values will materialize depending on the probability distribution of input variables. How to define project value will be discussed in later chapters. For the time being, suffice it to say that to be valuable and sustainable, a project must satisfy financial, environmental, and social objectives that are defined by management, shareholders, and other stakeholders.

It is not sufficient for a mining company to determine whether an asset has sufficient value to justify development of a new mine. It is also necessary to have an understanding of the uncertainty associated with variables that determine the project value, how this uncertainty creates risk, and how to manage this risk. The term *risk factors* or *risk variables* are used to define variables that create uncertainty in project evaluation.

To understand and manage the risk associated with evaluation of mineral assets, one must satisfy the following conditions:

- Understand the life cycle of a successful project. This life cycle can be divided into several stages. Decisions must be made after completion of each stage.
- Establish which information is needed to make sound decisions at the end of each stage. This information will be described as input variables, whose value is uncertain and may change in space or over time.
- Determine the range of values that these input variables might take and the probability that these values will materialize.

- Quantify the consequences that changes in the input variables will have on the probability that the company objectives will be achieved. Negative consequences include rejecting a project that would have a positive value if developed (false negative) or developing a project that will fail (false positive).
- Identify ways to reduce the probability of negative consequences while increasing the probability of achieving or exceeding specified objectives.

The types of errors that can be made when evaluating a project are represented in Figure 5.1. The horizontal axis shows the value at which a project is estimated. The vertical axis shows the true value of the project, which is unknown at the time decisions are made and which will only be known at the end of the mine life.

The graph is divided into five sectors, which are alphabetically defined as follows:

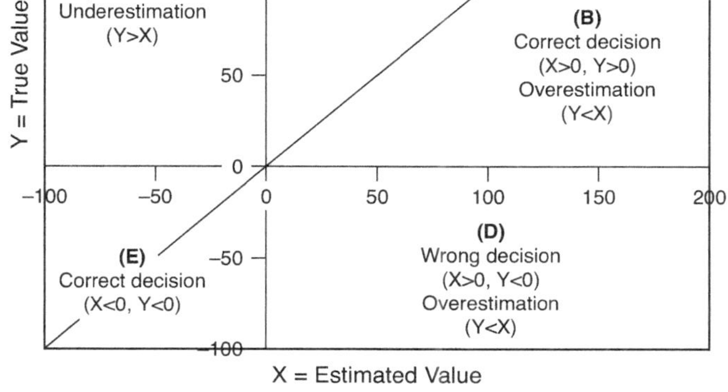

FIGURE 5.1 Graphical representation of the consequences of possible project estimation errors

- **Sector A:** The estimated value is positive ($X>0$). The decision is to go ahead with the project. The true value is also positive ($Y>0$), which confirms that the right decision was made. However, the value of the project was underestimated ($Y>X$).
- **Sector B:** The estimated value is positive ($X>0$). The decision is to go ahead with the project. The true value is also positive ($Y>0$), which confirms that the right decision was made. However, the value of the project was overestimated ($Y<X$).
- **Sector C:** The estimated value is negative ($X<0$). The decision is to stop the project. The true value is positive ($Y>0$), which means that a wrong decision was made (false negative). The value of the project was underestimated ($Y>X$).
- **Sector D:** The estimated value is positive ($X>0$). The decision is to go ahead with the project. The true value is negative ($Y<0$), which means that a wrong decision was made (false positive). The value of the project was overestimated ($Y<X$).
- **Sector E:** The estimated value of the project is negative ($X<0$) and so is the true value ($Y<0$). The correct decision is made, which is to stop a project with negative value.

Ideally one would want the estimated value to be as close as possible to the true value. This would require that no error be made, an objective that cannot be reached for practical reasons (excessive cost of risk reduction or impossibility to know future value of risk variables such as commodity prices). The primary objective should be to maximize the probability that the correct decisions are made: When a project is developed, its true value is positive (Sectors A and B), and when a project is stopped, its true value would have been negative if developed (Sector E). Simultaneously, the probability of false negative (Sector C) or false positive (Sector D) should be minimized.

LIFE-OF-MINE CYCLE REPRESENTED AS A LINEAR PROCESS

While the general approach to decision-making under uncertainty is independent of the project to be evaluated, each project has its own peculiarities. Evaluation of a mining project consists of developing a model of what the life

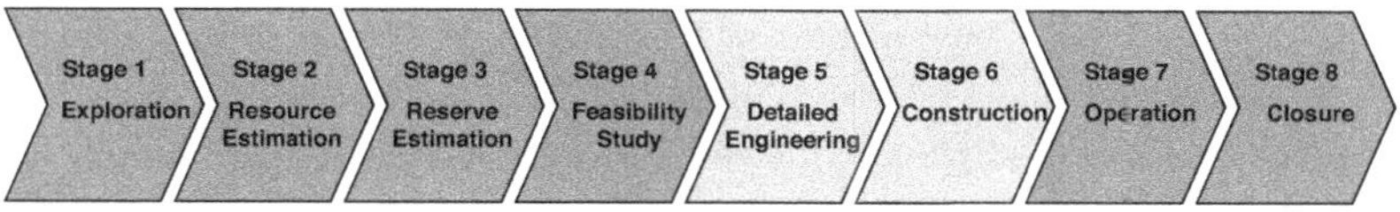

FIGURE 5.2 The mine life cycle as a linear process

of a mine would look like, both technically and economically, if such a mine were to be developed. The stages that characterize a typical project can be summarized as shown in Figure 5.2.

The life of a project is comprised of eight stages:

- **Stage 1—Exploration:** First, a mineral deposit must be discovered. If the information gathered is of sufficient interest, it might be classified and publicly reported as "exploration results" under international reporting rules, with mention of the exploration potential. A successful Stage 1 exploration program is that which reaches the point where a deposit model can be developed and form the basis of a scoping study.
- **Stage 2—Resource estimation:** The next stage consists of obtaining enough information to demonstrate that the mineral deposit presents the potential for eventual economic extraction. To reach the point where a mineral resource can be publicly disclosed, one must not only get more data than in Stage 1 but also develop a deposit model and complete high-level technical and economic scoping studies. Resources must be categorized as measured, indicated, or inferred.
- **Stage 3—Reserve estimation:** A reserve is the economically minable part of a measured and indicated resource. To estimate a mineral reserve, one must first complete a pre-feasibility study, which consists of analyzing several possible development options and choosing a preferred option. If the pre-feasibility study gives positive results, the preferred option becomes the basis for a feasibility study. Depending on circumstances, it may be possible to report a reserve on the basis of a pre-feasibility study, or a feasibility study may be required.
- **Stage 4—Feasibility study:** A comprehensive technical and economic study of the preferred option is completed with sufficient

detail that a final decision can be made concerning development. The results must be such that all key parameters are clearly defined as needed to sufficiently narrow the scope of detailed engineering design. Depending on the size of the project, feasibility studies are completed either internally by the owner or subcontracted to an engineering firm with owner representation.

- **Stage 5—Detailed engineering:** For large projects, this stage is usually subcontracted to an engineering, procurement, and construction (EPC) firm. Following successful completion of this stage, construction can be approved.
- **Stage 6—Construction:** The project infrastructure is built following specifications developed during detailed engineering. Construction is usually divided between activities completed by the owner and others completed by the EPC firm and its subcontractors. Completion tests must be passed before the mining company accepts the project as built. Project financing may also be conditional on successful completion of tests concerning initial operational performance.
- **Stage 7—Operation:** The mine operates and brings to market a salable product for as long as conditions are economically favorable and sufficient minable reserves exist. Factors that may prematurely terminate the profitable life of a project can be of a geologic, economic, environmental, social, or political nature.
- **Stage 8—Closure:** Environmental and social remediation must be completed to maintain a sustainable environment after the mine no longer operates. Appropriate project closure plays an important role in ensuring that the operator retains its public license to operate not only locally but worldwide.

Each one of these eight stages can be subdivided into several steps:

- **Step 1—Scope definition:** The first part of each stage consists of defining what the objective is and how this objective will be reached.
- **Step 2—Information gathering:** This step consists of getting the information needed to reach the objective. Information can be obtained by a variety of means such as drilling, sampling, testing, baseline surveys, market studies, or other means as appropriate for each stage.

- **Step 3—Analysis:** The data gathered during Step 2 must be analyzed. Analyses can be objective, based on measurements and quantitative studies, or subjective, based on experience and opinions. In most situations, both subjective and objective analyses are needed. Analyses can also be either deterministic or probabilistic. The methods to be used depend on the information available and the decision to be made.
- **Step 4—Decision:** After completing an analysis of the information, a decision must be made concerning what to do next. Decisions that might be made at the end of each stage include the following:
  - Go to the next stage.
  - Abandon the project, place it on a care-and-maintenance program, or sell it.
  - Remain in the same stage. Additional information is needed to decide whether to go to the next stage or abandon the project.

In this book we are primarily interested in the first four stages in Figure 5.2. These stages are critical in estimating the project value and controlling the risk associated with deciding whether or not to go ahead with development of a new mine. The costs associated with the first four stages are low compared with those incurred in the following four. Conversely, the time needed to reduce the risk to an acceptable level might be long compared with the time needed for detailed engineering and construction. If a project reaches the feasibility level (Stage 4), the objective is to evaluate the likely value of project development, which requires predicting what will be the outcome of the next four stages, from detailed engineering to mine closure.

MINE LIFE CYCLE WITH MULTIPLE DECISION POINTS AND MULTIPLE FEEDBACKS

A project life cycle is typically never so simple that it can be represented as a linear sequence of events as shown in Figure 5.2. It is better represented by Figure 5.3 where constant feedback is necessary to eventually reach a realistic evaluation. For example, mining and processing methods are assessed on the basis of available geologic information, but the geologic information that is needed to develop a life-of-mine production schedule is a function of

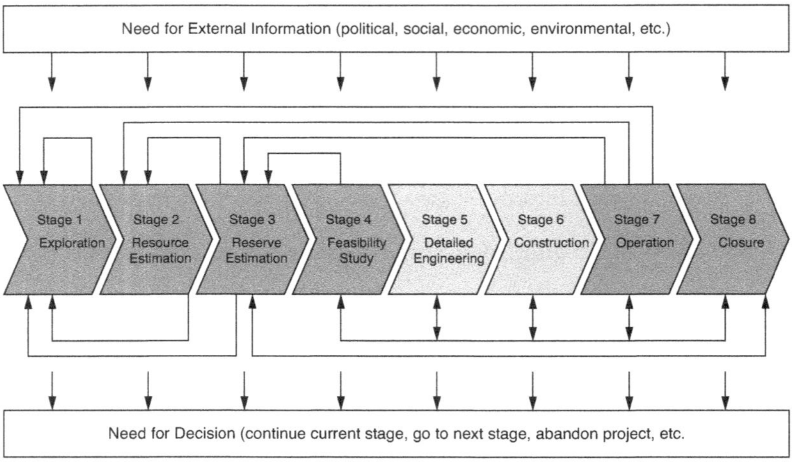

FIGURE 5.3 The minelife cycle with multiple decision points and multiple feedbacks

mining and processing methods. The value of the product to be sold, such as a copper-zinc concentrate, is determined by contractual agreements that specify acceptable levels of contaminants; knowing which levels are acceptable will influence the resource estimation process. A feasibility study may show insufficient information concerning structural geology, which will have to be obtained before infrastructure design can be finalized.

Figure 5.3 shows only the most obvious relationships between stages. Single arrows are used to show that completion of one stage may indicate the need to reexamine any of the previous stages. Stage 8 (closure) is dependent on decisions made during all previous stages. The double arrows that end on Stage 8 are there to indicate that the closure plan must be included in the feasibility study, and that project design, mine operation, and closure costs are strongly related.

CLASSIFICATION OF RISK VARIABLES AS TIME OR SPACE RELATED

To assess the risk associated with a given project, one must complete the following tasks:

- Develop predictive models that show the relationship between input variables and project value.
- Determine which are the input variables that could significantly influence the project value and how the value of one variable is dependent on or influences that of another.
- Assess the values which the input variables might take during the life of the project, the reason why these values are not known with precision, and the range or probability distribution of these values as they may change over time and space.
- Taking into account the joint probability distribution of the input variables and the relationship between their value and the value of the project, develop a model of the probability distribution of the project value.

Risk originates from incomplete or incorrect information concerning input variables. Risk also originates from incorrect interpretation of the relationship between input variables and project value. There are a number of reasons why input variables are not known with certainty. The first source of errors is in measurement (e.g., sampling and assaying errors). Another source of errors is in interpreting raw data and developing models to predict values either in space (e.g., deposit model) or over time (e.g., model of future metal prices). Finally, the relationship between the uncertainty concerning input variables and project risk may not be properly evaluated.

When analyzing the different sources of uncertainty in project evaluation, it is convenient to classify input variables as either space related (spatial) or time related (temporal). Both space- and time-related variables are subject to measurement and prediction errors. Different statistical methods can be used to model the level of confidence that can be ascribed to time- and space-related variables.

The geologic properties of a mineral deposit are clear examples of space-related variables. These properties do not change over time.[1] Surface observations,

1 This statement is valid only within the scope of a feasibility study. Once mining takes place, the properties of the remaining part of the deposit, such as geotechnical properties or degree of oxidation, may change. In situ leaching changes resource characteristics over large areas.

drill holes, and underground workings can be used to sample the deposit and to determine the value of these variables at specific locations. Errors are made when estimating values where the deposit is sampled, as well as when predicting values between observation points. Better information is obtained by increasing the number of sample points, but the cost and time needed to get this information can be significant, and eventually the benefit expected from more information no longer justifies the cost. Space-related variables include geology of the deposit, site geography, topography, preexisting infrastructure, and the preexisting social, environmental, and political environment.

The value of time-related variables is expected to change during the life of the project. With few exceptions, the timing and magnitude of these changes cannot be predicted with confidence. One cannot sample the future, but it might be possible to predict a number of equally probable realizations (outcomes) and to analyze the project sensitivity to such realizations. Falling in the category of time-related variables are the cost of inputs (consumables, labor, etc.), the value of output (price at which the final product will be sold), the timing and magnitude of natural events (weather patterns, earthquakes, hurricanes, etc.), and the social, environmental, and political situation, as it may change during the life of the project.

Figure 5.4 illustrates the relationships between risk variables depending on whether the lack of knowledge concerning their value results from possible changes in space (left circle) or over time (right circle). The circles are overlapping to show that some risks, such as those resulting from the environmental impact of a project, have both a spatial and temporal component. The characteristics of space- and time-related variables are summarized in Table 5.1.

In the next chapters, a variety of statistical methods are introduced that can be used to assess and manage risk, depending on whether the risk is space or time dependent.

CLASSIFICATION OF ESTIMATION ERRORS ACCORDING TO PRECISION AND ACCURACY

Errors are made when estimating the value of risk variables and using these values to evaluate a project. Data acquisition and data analysis are sources of

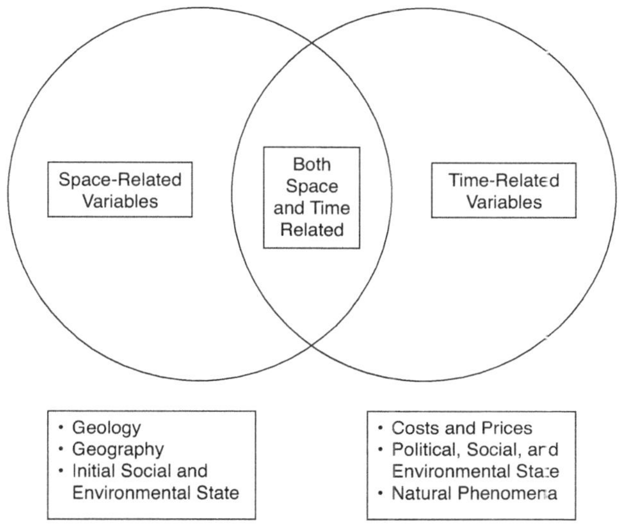

FIGURE 5.4 Classification of risk-generating variables as space or time related

estimation errors. If a sample is analyzed repeatedly in the same laboratory, the results are not likely to be the same every time, but they are expected to be centered around one value, which is the laboratory best estimate. However, if the same sample is analyzed in different laboratories, systematic differences are likely to be observed, some laboratories giving generally higher results while others give lower results. If the results obtained in one laboratory show little variability, they are said to be precise. If the results obtained in one laboratory differ significantly from those obtained in other laboratories, the results are said to lack accuracy. Precision is generally understood as the result of random errors, whereas accuracy results from systematic errors.

The difference between precision and accuracy is best explained by showing the results that can be obtained when shooting repeatedly at a target (Figure 5.5). If all shots are close to the center, the results are defined as both precise and accurate. If the shots are spread over a large area and centered away from the center, the results are both imprecise and inaccurate.

Long-term forecasts of future metal prices are likely to be both imprecise and inaccurate. These estimation errors can be reduced by market studies, but even the best studies have been unable to eliminate both random and systematic

TABLE 5.1 Comparison of space-related and time-related variables

	Type of Risk Variables	
	Space Related	**Time Related**
Space/time relationship	Value varies in space and is fixed over time.	Value varies over time and is independent of location.
Source of uncertainty	Value is observed at specific locations but is unknown between observation points. Magnitude of uncertainty increases with distance between observation points or extrapolation beyond observation points.	Current value is known; future value is unknown. Magnitude of uncertainty generally increases with time.
Management of uncertainty	Additional spatial information can be obtained to reduce uncertainty. Simulation can be used to quantify uncertainty between observation points.	Future values cannot be determined with certainty, but the timing and magnitude of future changes in value can be simulated.
Impact on project	Project performance is highly dependent on value of both space- and time-related variables. Project design must include built-in flexibility to adapt to future changes in input variables.	

errors (over- or underestimation). Before accepting a prediction from a specific source, one should analyze the performance of earlier predictions. Some individuals or companies tend to predict that the last few years are indicative of the years ahead: downward trends are projected downward while upward trends are projected upward. Others are eternally optimistic, systematically predicting higher prices, especially during periods of depressed prices.

Precision and accuracy of deposit models are a function of the quality and quantity of the data used to develop the models and of the modeling method itself. Different data interpretations and different modeling methods will give different results that are all likely to differ from the actual state of the world being represented.

A deposit model can be precise in the sense that it predicts the location of high- and low-grade areas accurately, while being inaccurate in that the grade of predicted high-grade areas is overestimated and that of predicted low-grade areas is underestimated. Even the best deposit models are likely to give results that are both accurate and precise only when large tonnages are considered, while lacking accuracy when being used to predict mine production over short time periods. Precision and accuracy of estimates can be improved by increasing the quantity and quality of data and by ensuring that the persons responsible to complete the estimation are qualified in both the theoretical

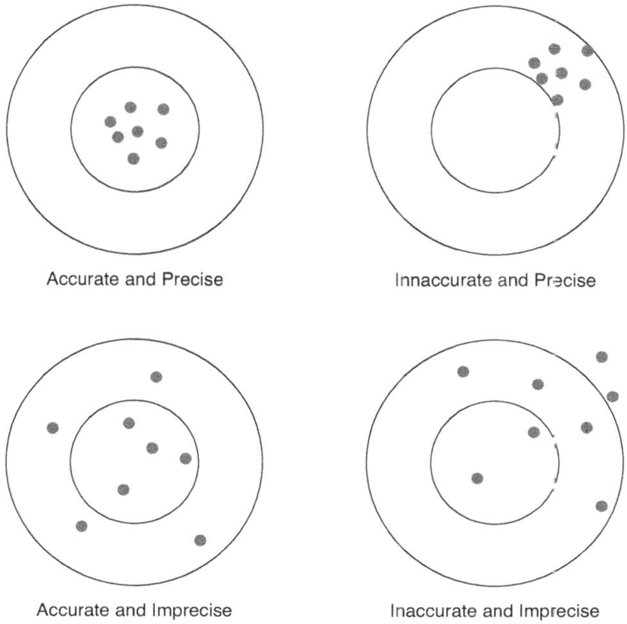

FIGURE 5.5 Graphical representation of accuracy and precision

and practical aspects of resource estimation. In most circumstances, there are both technological and economic limits to the precision and accuracy of estimates that can be achieved in a timely fashion and at a reasonable cost.

Precision and accuracy can be used to characterize not only risk variables but also estimates that are derived from these variables. The possible effect of lack of precision and accuracy must be taken into account when estimating a project risk. A project capital cost that is estimated at $157 million, plus or minus 5%, is extremely precise and might be interpreted as being accurate. But if the actual cost of construction is $196 million, or 25% higher than predicted, this precise estimate is proven (after the fact) to have been highly inaccurate. Cost underestimation is a common cause of project underperformance or failure.

When estimating the value of a project, high precision and high accuracy may be considered as desirable objectives. However, there are practical limits to how precise and accurate an estimate can and should be. Risk reduction takes time and can only be achieved at a cost. Uncertainty always remains concerning the unknowable future value of time-related variables. The consequences of lack of either precision or accuracy vary with the project, and the need for such precision or accuracy varies accordingly. This is best explained by considering Figure 5.1. A point on the diagonal (Y = X) represents a project whose estimated value is both precise and accurate. In practice, no project can be estimated with perfect precision and accuracy. Within limits,[2] estimation of a high-value project (the true value Y is unknown but likely to be very high) need not be extremely precise or accurate, provided the errors made do not change the decision to go ahead (the estimated value X is neither accurate nor precise, but it is significantly positive). For this reason, most companies will tend to accelerate development of highly valuable projects, recognizing that the probability of a false positive is very low.

There are, however, many examples of situations where projects were estimated as having a high positive value but were proven later to be uneconomic. In some cases, the reason was unforeseeable changes in the value of critical risk variables. But in many cases it was because the estimated value was presented to management as highly positive and both accurate and precise. The estimate might have been accurate, but lack of precision meant that the probability of failure (false positive) was significantly higher than predicted.

Similarly, projects that have a negligible likelihood of having a positive value will be quickly rejected even if the estimates are based on a scoping study that is neither accurate nor precise: The probability of a false negative is negligible.

Where precision and accuracy are most relevant is for marginal projects (the estimated value X is near zero). Lack of either precision or accuracy is likely to result in a false positive or false negative. Consequently, marginal projects tend to be subject to longer and more detailed studies before a final decision is made. A decision might be to postpone a marginal project until conditions change, either positively or negatively, and the risk of making the wrong decision is considered acceptable.

2 Even for high-value projects, precision and accuracy must be reduced to the point where not only the decision to go ahead can be made with confidence but other critical decisions that influence future value can also be made reliably. Over- or undersizing the processing plant may not prevent the project from being economic but may significantly reduce its economic value. Appropriate sizing may require geologic information beyond that needed to justify a positive development decision.

CHAPTER 6

RISK ASSESSMENT USING MONTE CARLO SIMULATION

In the previous chapter, risk factors were classified according to whether lack of knowledge resulted from variability in space or over time. When variability is spatial, additional information can be obtained before deciding whether to proceed with the project. However, there is a cost to getting this data, and there is a point where this cost exceeds the value of additional information. Confidence decreases when projections are made farther away from known data points. When variability is temporal, one can estimate with various levels of confidence the value that the variable might take at a specific time. Such prediction is usually based on recent historic behavior and on related economic trends. Confidence progressively decreases when projections are made further into the future from the current time.

In this and the next chapter, two statistical methods are presented that can be used to make decisions under uncertainty: Monte Carlo simulation and decision tree analysis. Simple examples are given to familiarize the reader with the concepts and the advantages and disadvantages of using these methods. Complexity increases significantly when dealing with multistage projects whose value depends on a wide variety of intercorrelated risk variables. The uncertainty of these variables can be spatial or temporal, and their influence on the bottom line can be difficult to model.

Many books have been published on the theory of statistical decision-making under uncertainty. Analytical methods grouped under the term *real options analysis* are increasingly used. Consulting companies offer their services to mining companies for application of these methods. Many research centers are dedicated to the study of decision-making under uncertainty, either in a general context (DMDU n.d.; Rand Corporation 2016; IRM 2016; SRA 2016), or more specifically as it relates to the mining industry (COSMO n.d.). Discussing *real options analysis* is outside the scope of this book. One can argue that understanding Monte Carlo simulation and decision tree analysis is necessary before more complex methods can be applied intelligently.

GENERAL CONSIDERATIONS

When analyzing whether a new project should be developed, the mining company has specific objectives in mind. The value of a project is a function of the extent to which these objectives are reached. The expected dollar value of a project is not a perfect representation of what the project is worth to the company at the time the decision is to be made. Expected dollar values do not take into account the company's risk tolerance. Everything else being equal, two companies that consider the same project at the same time will give it different values depending on the extent to which they are risk averse, risk neutral, or risk loving. For these reasons, decision-making theorists have introduced the concept of "utility function." Risk tolerance and utility function are discussed in Chapter 9. The main component of the utility function is the dollar value of the project. While "utility" terminology and notations are used in this and the next chapters, no distinction is made here between utility and dollar value.

The value of a project is best represented by a function $U(x_1, x_2, ..., x_n)$ which defines how the utility of the project depends on a number of controlling input variables x_i, $i = 1...n$. In the simplest case, this utility is measured by the net present value (NPV) of the project, whose value depends on tonnage and grade of material expected to be mined and processed over time, metallurgical recovery, capital and operating costs, value of product sold at the time it is sold, and multiple other technical, financial, social, and environmental variables. The value of these variables is uncertain at the time the decision must be made, and the relationship between input variable and project utility may lack clarity.

The following information is needed to determine the probability distribution of the utility of developing the project, $U(x_1, x_2, ..., x_n)$:

- The formula that relates the project utility to the input variables. This formula is easily defined in mathematical terms when the utility is measured by such financial indicators as the NPV of future cash flows or the internal rate of return. It can, however, be significantly more complex when other intangible or "soft" factors come into question, such as social and environmental impacts, strategic objectives, ethics, and company values.
- The joint probability distributions of the input variables, including any cross-correlation between variables, where applicable. These probabilities can be objective or subjective.

Objective probabilities are quantified on the basis of available numerical information. For example, the three-dimensional probability distribution of the grade of a mineral deposit can be estimated by stochastic simulation using drill hole data as input. Subjective probabilities are based on judgment, and their reliability is dependent on experience and qualification of the individual responsible for their estimation. For example, projections concerning the possible range of future metal prices are mostly of a subjective nature.

Experience and qualification are fundamental requirements, whether the probabilities are objective or subjective. For example, many assumptions are made when simulating different possible realizations of the grade distribution in a mineral deposit. Whether these assumptions are valid depends in part on judgment. Uncertainty estimates should take into account not only the results of mathematical analyses but also the simplifying assumptions underlying most mathematical models, while giving due consideration to experience and qualification of estimator.

THE MONTE CARLO SIMULATION PROCESS

The Monte Carlo statistical method was developed in the 1940s to analyze processes whose outcome is uncertain. The method is aimed at determining the probability distribution of a specified outcome measured by the value of the utility function $U(x_1, x_2, ..., x_n)$. The Monte Carlo process consists of the following steps:

1. Possible values of the input variables x_i, i = 1...n are selected randomly from their probability distributions.
2. The value of the utility function is calculated based on these randomly selected values of the input variables. This value is defined as a realization.
3. Step 1 and Step 2 are repeated a number of times. Different values of the utility function are obtained from each simulation. In most cases, hundreds or thousands of Monte Carlo simulations are needed to obtain meaningful results.
4. The frequency distribution of the simulated values of the utility function is used to determine the risk associated with the project.

Monte Carlo simulation can be used to estimate the probability distribution of a project's NPV, including the probability of negative NPV, the likelihood that losses might exceed a specific threshold, or the probability of a very high return. The decision whether to proceed with the project can then be made taking these probabilities into account as well as the company's risk tolerance.

As an example, the estimated probability distribution of a project's NPV is shown in Table 6.1. From this distribution, one concludes that the project's expected NPV is US$145 million. However, there is 25% probability that the project will have a zero or negative NPV, and 10% probability that this loss will exceed $100 million. Depending on the decision maker's risk aversion, this risk of failure may be considered unacceptable. On the positive side, there is 20% probability that the NPV will be at least $300 million, an opportunity one may not want to lose. A decision may be made to acquire more information to reduce the downside risk to a reasonable level.

In the rest of this chapter, simple examples are given to illustrate the theory behind Monte Carlo simulation when considering input variables whose values are either independent of, or correlated with, those of other variables.

TABLE 6.1 Probability distribution of a project's net present value

Possible Outcome (NPV in millions)	Probability That Outcome Will Be Realized	Cumulative Probability Distribution
–$300	0%	0%
–$200	3%	3%
–$100	7%	10%
$0	15%	25%
$100	29%	54%
$200	26%	80%
$300	10%	90%
$400	5%	95%
$500	3%	98%
$600	2%	100%
$700	0%	100%
Weighted average	Total	
$145	100%	

MONTE CARLO SIMULATION OF UNCORRELATED VARIABLES

Consider a project that is designed to sell a product whose future value is highly uncertain. This value, expressed in U.S. dollars per metric ton, is expected to be between $10 and $65, the most likely range being $20 to $35. A market study was completed from which the probability distribution of future prices was estimated to be as shown in Table 6.2 and represented in Figure 6.1. Also shown in Table 6.2 is the cumulative distribution of future prices. It can be seen that there is 38% probability that the price will be less than or equal to $25, and 17% probability that it will be more than $40. The cumulative distribution is plotted in Figure 6.2. Note that the probabilities need to add up to 100% and the cumulative distributions can only have values between 0% and 100%.

The following steps must be followed to randomly pick a future price from the probability distribution and estimate the probability distribution of the project's NPV:

TABLE 6.2 Probability distribution of future price of product sold

Possible Outcome (per metric ton)	Probability	Cumulative Probability
$10	0%	0%
$15	5%	5%
$20	15%	20%
$25	18%	38%
$30	20%	58%
$35	15%	73%
$40	10%	83%
$45	7%	90%
$50	5%	95%
$55	3%	98%
$60	2%	100%
$65	0%	100%
Weighted average	Total	
$32.0	100%	

1. Generate a random number u(0,1) between 0 and 1 from a uniform distribution.[1,2] For this example, it is assumed that the generated random number is 0.71.
2. Determine the price of product sold that corresponds to the 71% cumulative probability distribution. Figure 6.3 shows that this price is x = $34 per metric ton.
3. The NPV U(x) is calculated using the $34 sales price.
4. The preceding steps are repeated 100 times. For each iteration, different values of x and U(x) are obtained. Each value has an equal probability of being realized.

1 A uniform distribution is one in which all numbers between 0 and 1 have an equal likelihood of occurrence. There are various types of random number generators, one of them being the Microsoft Excel RAND() function.

2 The utility function U(x) is capitalized. The random number u(0,1) is not.

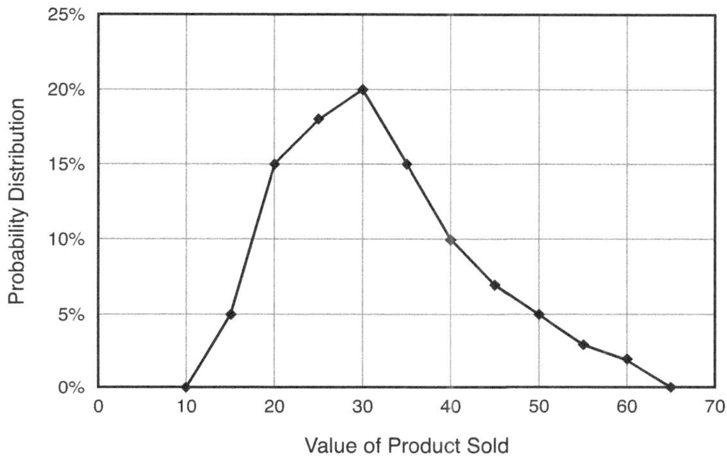

FIGURE 6.1 Probability distribution of future price of product sold

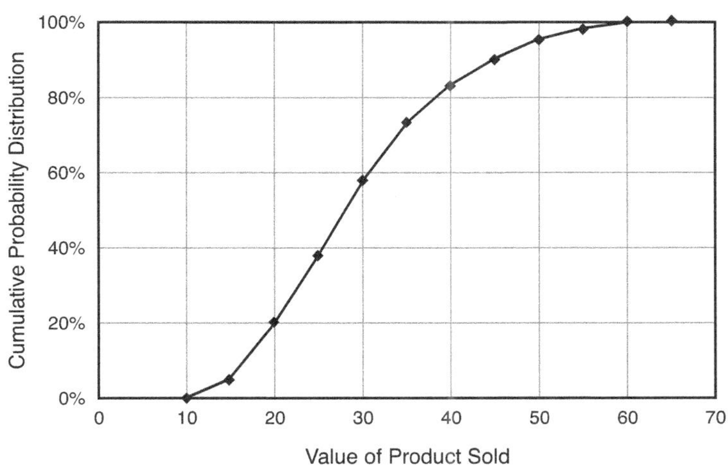

FIGURE 6.2 Cumulative probability distribution of future price of product sold

5. Compare the probability distribution of the 100 simulated values of x with the input distribution shown in Figure 6.1. If the two distributions are significantly different, the number of iterations should be increased. Probability distributions obtained after 25, 50, 75, and 100 simulations are shown in Figure 6.4. In most situations, more than 100 simulations are needed to reasonably reproduce the input probability distribution.
6. Calculate and plot the frequency distribution of the 100 (or more) simulated values U(x). From this frequency distribution, one can answer some important questions, such as what is the probability of failure ($U(x) < 0$), or what is the probability that the project value will exceed some threshold U_{max} ($U(x) > U_{max}$).

If the project utility is a function $U(x_1, x_2, ..., x_n)$ of multiple variables that are uncorrelated, random values of x_i, $i = 1, ..., n$ must be generated independently. A different random number $u_i(0,1)$ must be used for each variable x_i. These values are used to calculate the corresponding utility $U(x_1, x_2, ..., x_n)$. The process is repeated 100 times, thus obtaining 100 possible utilities. The number of iterations should be increased when the number of variables increases.

MONTE CARLO SIMULATION OF MULTIPLE CORRELATED VARIABLES

If two or more input variables are correlated, the applicable Monte Carlo method is similar but slightly more complex than that just described for uncorrelated variables. For example, consider the utility function $U(x_1, x_2)$ of a copper-gold project whose value is a function of the copper grade x_1 and the gold grade x_2. Gold grade and copper grade are correlated, but this correlation is not perfect. (If two variables are perfectly correlated, they can be treated as a single variable.) The joint probability distribution of x_1 and x_2, $p(x_1, x_2)$, can be estimated from drill-hole information or from a computer-generated deposit model. This probability distribution specifies not only the probability distribution $p(x_1)$ of the copper grade x_1 but also the probability distribution $p(x_2|x_1)$ of the gold grade x_2 that corresponds to a copper grade x_1. These probabilities are related as follows:

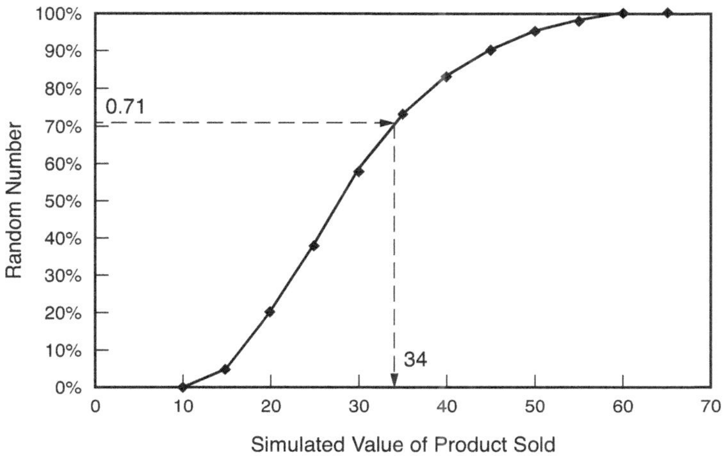

FIGURE 6.3 Random selection of a future price

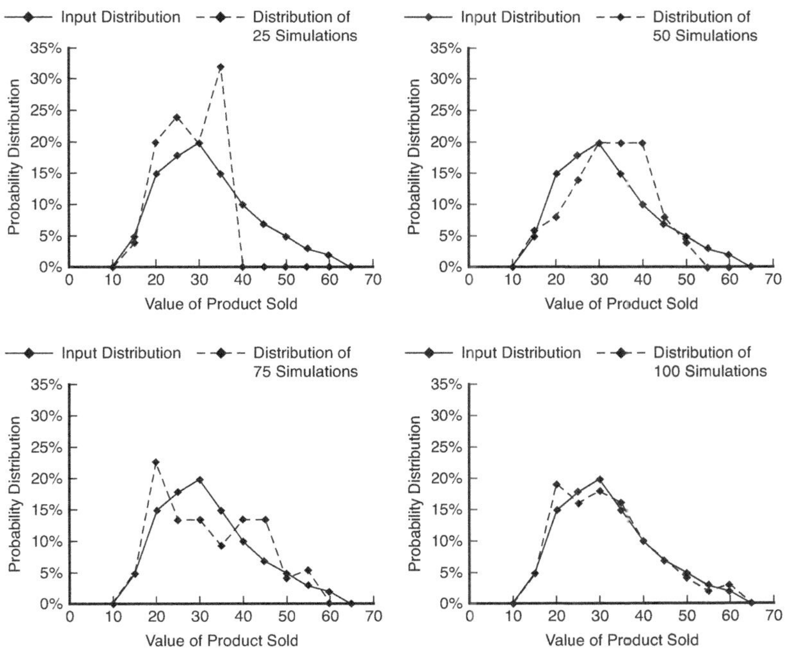

FIGURE 6.4 Comparison of 25, 50, 75, and 100 simulations of probable values of product sold

$$p(x_1, x_2) = p(x_2|x_1)\ p(x_1)$$

To randomly generate a single value of the utility function, one must generate a random value of x_1 using the probability distribution $p(x_1)$ and a random value of x_2 using the probability distribution $p(x_2|x_1)$ in which x_1 is known. In both cases, the random-value generation process is reduced to the single-variable process described in the previous section. Appendix A illustrates how to randomly generate a realization of the average grades of a copper-gold deposit when copper and gold are correlated.

This two-variable approach can be applied to any number of variables. For example, if the utility of a project depends on the value of three correlated variables, the relationship between these variables can be written as follows:

$$p(x_1, x_2, x_3) = p(x_3|x_1, x_2)\ p(x_2|x_1)\ p(x_1)$$

The process to be followed to randomly generate a set of correlated values $\{x_1, x_2, x_3\}$ is reduced to three successive single-variable random number generations:

1. Randomly generate a value of x_1 using $p(x_1)$.
2. Randomly generate a value of x_2 using $p(x_2|x_1)$, where x_1 is known.
3. Randomly generate a value of x_3 using $p(x_3|x_1,x_2)$, where x_1 and x_2 are known.

There are many circumstances when the data necessary to estimate complex joint probability distributions such as $p(x_1, x_2, x_3)$ are not available and approximations must be made. For this reason, simplifying assumptions are often made in commercial software designed to implement Monte Carlo processes. For example, when two variables are considered x_1, x_2, the necessary input is reduced to the one-variable probabilities $p(x_1)$ and $p(x_2)$, and the correlation coefficient $r(x_1,x_2)$ between x_1 and x_2 whose value must be between −1 (perfect negative correlation) and +1 (perfect positive correlation). Conversely, complex methods, such as stepwise conditional transformation, have also been developed to simulate multiple variables (Leuangthong and Deutsch 2003).

MONTE CARLO SIMULATION AND EVALUATION OF STAGED PROJECTS

Monte Carlo simulation can be used to simulate realizations of the values that risk variables $x_1, x_2, \ldots, x_n$ might take. The project utility $U(x_1, x_2, \ldots, x_n)$ is calculated for each simulation. The resulting probability distribution gives valuable information concerning project risk and opportunities. Mining companies routinely use Monte Carlo simulation to evaluate projects and decide the path forward. This approach is, however, not immediately applicable when considering the sequential aspect of project evaluation, which comprises multiple stages, from exploration to project development. Each stage is designed to achieve specific objectives that are a function of the results obtained in previous stages. The decision to go ahead with a given stage is a function of the probability that this stage will reach its objectives. Chapter 7 shows how decision trees, combined or not with Monte Carlo simulation, can be used to effectively model multistage projects.

CHAPTER 7

DECISION TREE TO EVALUATE MULTISTAGE PROJECTS

Project evaluation consists of the completion of a number of stages (Chapter 5). Each stage is designed to decrease the probability of making the wrong decision concerning development of a new mine. Two possible outcomes have to be considered, including rejection of a project of economic value (false negative) or development of an uneconomic mine (false positive). The probability of occurrence of these outcomes must be reduced to an acceptable level. In early stages, the risk is high, but the cost of risk reduction is low. With successive stages, the remaining risk is less, but the cost of further reduction increases. These costs, which increase geometrically, can only be justified if completion of the corresponding stage is expected to result in commensurate reduction in the project risk. Decision tree analysis is an effective way to graphically represent and quantitatively evaluate a multistage decision process.

GRAPHICAL REPRESENTATION OF A MULTISTAGE PROJECT

One of the major sources of uncertainty originates from a limited understanding of the geologic properties of the mineral deposit. The risk resulting from this limited knowledge can be reduced by getting more or better information, for example, by infill drilling, drill hole logging, geochemical or geometallurgical testing, mapping, deposit modeling, or other means.

A useful way to represent and evaluate a proposed exploration project is with a decision tree such as the one shown in Figure 7.1. In this example, a

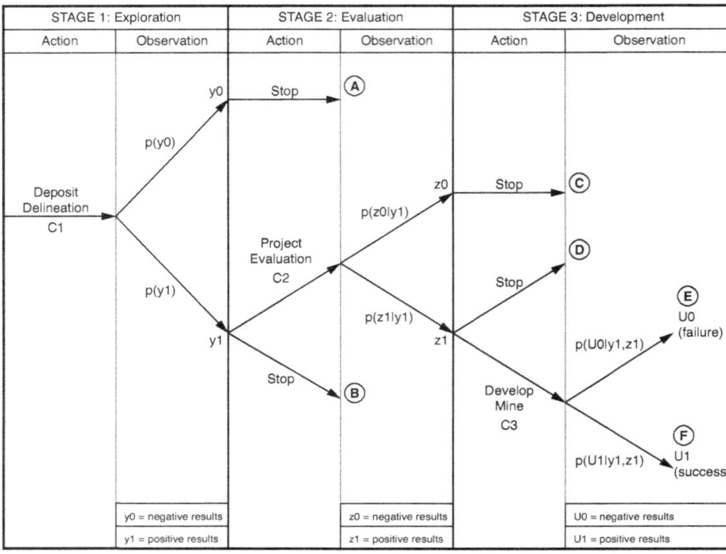

FIGURE 7.1 Decision tree example when risk is space related

geological anomaly has been identified and a decision must be made whether drilling is justified to characterize the underlying mineralization. If the results are positive, a feasibility study will be completed, which may lead to the development of a new mine.

The process to be followed has been divided into three stages: exploration, evaluation, and development. Each stage is further divided into two parts: action and observation. The decision to act in a specific way is made on the basis of available information. There is a cost, C, associated with each possible action. Following this action, additional information is obtained. The observation can be positive, to the extent that it supports moving into the next stage, or negative, which in this example results in abandoning the project. The observation made at the end of Stage 1 is defined by the letter y, with index 1 ($y = y_1$) for a positive result and 0 ($y = y_0$) for a negative result. The letters z and U define the observations made at the end of Stage 2 and Stage 3, respectively. If the decision is made at the end of Stage 3 to develop a mine, the observation that will be made is that the project has a positive ($U = U_1$) or negative ($U = U_0$) value. The expected project value U is often (but not necessarily correctly[1]) measured by the net present value of expected future cash flows. To decide which action to take, one must estimate not only the cost of this action but also the likelihood that acting will result in positive or negative results.

The stages shown in Figure 7.1 can be described as follows:

- **Stage 1, Exploration.** This stage consists of drilling and analyzing the results obtained. These results, defined by the variable y, can be positive ($y = y_1$) or negative ($y = y_0$). To justify the cost C_1 of exploration, the geologist must have a subjective estimate of the probability $p(y_1)$ of obtaining positive results. The probability of negative results is $p(y_0) = 1 - p(y_1)$.
- **Stage 2, Evaluation.** If the exploration results are negative, the project is stopped. If positive results are obtained, a feasibility study will be completed. The cost of this study is C_2. The results z can be positive ($z = z_1$) or negative ($z = z_0$). Given that exploration was successful ($y = y_1$), the probability of the feasibility study giving positive results is $p(z_1|y_1)$. This probability is a function of not only geology but also all other modifying factors. It must be estimated by the project manager, taking into account the opinion of geologist, mining and metallurgical engineers, cost estimators, and so on.
- **Stage 3, Development.** If the result of the feasibility study is positive ($z = z_1$), the decision will be to develop the project. The cost of development is C_3. The expected value of the mine operation is measured by its utility, U. It is recognized that even if exploration was successful ($y = y_1$) and the feasibility study gave positive results ($z = z_1$), there remains some risk, however small, that the project will fail. The probability that the project will be successful ($U = U_1$) is $p(U_1|y_1, z_1)$, which ideally should be equal to 100% but is likely to be less.

Funds that are spent in Stages 1 and 2 are aimed at reaching a number of conflicting objectives:

1 The expected net present value is only one possible measure of a project's utility. This is further discussed in Chapter 9.

- Keeping the cost and duration of exploration C_1 and feasibility study C_2 to a minimum
- Increasing the probability $p(U_1|y_1, z_1)$ of developing a successful project
- Decreasing the probability $p(U_0|y_1, z_1)$ of developing a failing project
- Minimizing the probability $p(y_0|U_1)$ of abandoning the project after Stage 1 ($y = y_0$) when the value of the project to be discovered is positive ($U = U_1$)
- Minimizing the probability $p(z_0|U_1)$ of abandoning the project after Stage 2 ($z = z_0$) when the value of the project to be discovered is positive ($U = U_1$)

USE OF A DECISION TREE TO ANALYZE AN EXPLORATION PROJECT

The decision tree shown in Figure 7.1 can be used to decide whether the proposed exploration program should be completed. To answer this question, one needs additional information.

- **Geologic potential:** The geologist should state the type of deposit (exploration target) that the area being explored is likely to contain. With assistance from mining engineers, metallurgists, and project engineers, a dollar value U_1 should be given to this type of deposit. The probability that there exists such a deposit is called the "prior" probability $p'(U_1)$. This probability is only dependent on geology and is independent of whether a decision is made to explore the area.
- **Exploration program:** The effectiveness of the exploration program is measured by the probability that it will lead to successful development of an economic mining operation with value U_1. This probability, $p(y_1,z_1,U_1)$, depends on all three stages being successful ($y = y_1$, $z = z_1$, and $U = U_1$) and cannot be greater than $p'(U_1)$. Even if there is a deposit of economic value in the area of interest, there is no guarantee that the exploration program will discover it ($y = y_1$)

and that the feasibility study will demonstrate that it can be developed economically ($z = z_1$).

- **Effectiveness of Stage 1:** There is a cost C_1 that will be incurred in completing exploration as planned in Stage 1. The geologist needs to estimate the probability $p(y_1)$ that Stage 1 will give positive results that will justify going to Stage 2. This probability depends on both geology and the proposed drilling program.
- **Effectiveness of Stage 2:** The cost C_2 must be estimated for the Stage 2 feasibility study that may follow successful exploration. The cost of the feasibility study will depend on the amount of information obtained during Stage 1 exploration. The probability $p(z_1|y_1)$ that the feasibility study will give positive results ($z = z_1$) following successful exploration ($y = y_1$) must also be estimated.
- **Effectiveness of Stage 3:** If the decision is made to develop a new mine, the capital cost C_3 of such a decision must be estimated. It is a function of the location of the project and the type of deposit likely to be discovered. The observation that will result from this decision can be measured by the likely net present value (NPV) of the cash flow that the mine will generate if it is built. In this example, it is assumed that there are only two possible outcomes: the project is successful ($U = U_1$) and its NPV is positive, or the project fails ($U = U_0$) and its NPV is negative. The magnitude of these positive or negative NPVs must be estimated.

Monte Carlo simulation can be used to estimate the probability that a discovered deposit will have a positive or negative value if developed. The process will give not only the probability of two values ($U = U_0$ and $U = U_1$) but also the probability distribution of a range of values. To estimate these probabilities, one must consider the expected geological properties of the exploration target and the uncertainty that will remain after completion of a feasibility study. This uncertainty must take into account not only geology but also financial, social, and environmental factors that will influence the project's NPV but are imperfectly known at the time that the decisions are made. To decide whether or not exploration should take place, the cost of each future stage, the possible value of a discovery, and the probabilities of success or failure must be estimated with limited or no physical information. Probabilities are most likely to be subjective.

A new mine will be developed only if both Stage 1 and Stage 2 are successful. The probability that the proposed exploration project will result in development of a new mine is $p(y_1,z_1,U_1)$. As shown in Chapter 6, this probability can be decomposed as follows:

$$p(y_1,z_1,U_1) = p(U_1|\, y_1,z_1)\, p(z_1|y_1)\, p(y_1)$$

Another useful relationship between probabilities is as follows:

$$p(y_1,z_1,U_1) = p(y_1,z_1|U_1)\, p'(U_1)$$

These two equations show that there are different ways to estimate the probability of success. In the first equation, the probability of each stage must be estimated assuming that the previous stage was successful. In the second equation, the probabilities that must be estimated include the probability that there is a deposit of economic value in the area of interest and the probability that this deposit will be discovered. Each equation can be used to estimate $p(y_1,z_1,U_1)$. The results should be the same. However, the intellectual process to be followed by the geologist to estimate each component probability differs significantly.

The estimated probabilities should be independent of the method used to estimate them. This remark applies to all probabilities shown in Figure 7.1. A recommended process is to estimate separately a number of probabilities and then ensure that they are consistent with each other. In the example that follows, we show that only four probabilities are needed, from which all other probabilities can be calculated.

NUMERICAL ANALYSIS OF AN EXPLORATION PROJECT

Table 7.1 summarizes information that is needed to analyze the value of the exploration project depicted in Figure 7.1. This information includes not only probabilities but also costs and likely outcome, depending on whether the project is developed and economically successful. The project's expected NPV is exclusive of the capital cost C_3. All values are subjective estimates made by those responsible for the project using best available data and expertise gained from experience.

TABLE 7.1 Probabilities, costs, and project utility needed to justify the exploration program

Notation	Definition	Value
Probabilities (estimated by geologists and project engineers)		
$p'(U_1)$	Probability that there exists a deposit of value U_1	10%
$p(y_1)$	Probability that Stage 1 will give positive results	30%
$p(z_1\|y_1)$	Probability that Stage 2 will give positive results if Stage 1 is positive	40%
$p(y_1,z_1,U_1)$	Probability that a deposit of value U_1 will be discovered	8%
Cost of three stages ($ millions, discounted to present)		
C_1 =	Cost of Stage 1, exploration	–$10
C_2 =	Cost of Stage 2, feasibility	–$70
C_3 =	Cost of Stage 3, mine development	–$300
Net present value of project cash flow after development ($ millions)		
U_0	NPV of project if developed and it fails (does not include C_3)	–$200
U_1	NPV of project if developed and it succeeds (does not include C_3)	$1,000

From the estimates shown in Table 7.1, a number of relevant probabilities can be calculated whose values should be reviewed for reasonableness:

- Probability that both Stage 1 and Stage 2 will be successful:

$$p(z_1,y_1) = p(z_1|y_1)\ p(y_1) = 40\% \times 30\% = 12\%$$

The probability that the exploration project will be abandoned, either at the end of Stage 1 ($p(y_0) = 70\%$) or at the end of Stage 2 ($p(z_0,y_1) = p(z_0|y_1)\ p(y_1) = 60\% \times 30\% = 18\%$), is 88%.

- Probability that a project of economic value will be developed after successful completion of Stage 1 and Stage 2:

$$p(U_1|y_1,z_1) = p(y_1,z_1,U_1) \ / \ p(y_1,z_1) = 8\% \ / \ 12\% = 66.7\%$$

The probability that a new mine will fail is $p(U_0|y_1,z_1) = 1 - p(U_1|y_1,z_1) = 33.3\%$. There is one chance in three that, if the decision is made to develop a new mine, the net present value of future cash flows will be negative.

- Probability that both Stage 1 and Stage 2 will give positive results when a deposit of economic value is present:

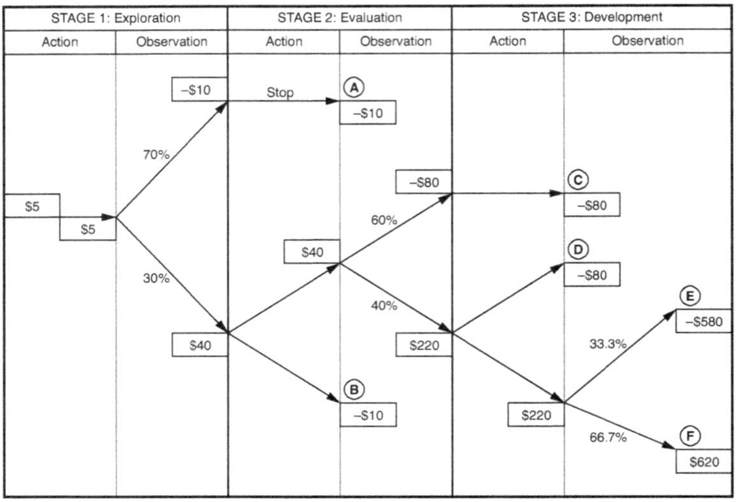

FIGURE 7.2 How to use the decision tree to evaluate the best exploration decision

$$p(y_1,z_1|U_1) = p(y_1,z_1,U_1) \,/\, p'(U_1) = 8\% \,/\, 10\% = 80\%$$

In other words, if there is a deposit in the area of interest that has a positive expected value U_1, there is 80% probability that Stage 1 and Stage 2 will give positive results ($y = y_1$ and $z = z_1$) and that the decision will be made to develop it. This may sound as reasonable odds. However, there is 20% probability that Stage 1 and Stage 2 will give positive results even if the project has a negative NPV.

- Probability that both Stage 1 and Stage 2 will give positive results when a deposit of negative NPV is present:

$$p(y_1,z_1|U_0) = 1 - p(y_1,z_1|U_1) = 20\%$$

The probabilities that correspond to each possible observation are shown in Figure 7.2. Also shown in Figure 7.2 is the cost or benefit (utility) of reaching the possible final decisions, which are marked by the letters A to F at the end of each tree branch. Costs are incurred to reach all decisions. In the event that A, B, C, or D is reached, the project is not developed. If E is reached, the project is developed but generates a negative cash flow. F is only reached if all exploration, feasibility, and operation results are positive. In this situation, the cash flow generated is sufficient to cover all costs and results in a positive \$620-million project NPV.

The utilities associated with each possible final decision are calculated as follows:

$$U(A) = -C_1 = -\$10$$

$$U(B) = -C_1 = -\$10$$

$$U(C) = -C_1 - C_2 = -\$80$$

$$U(D) = -C_1 - C_2 = -\$80$$

$$U(E) = -C_1 - C_2 - C_3 - U_0 = -\$380 - \$200 = -\$580$$

$$U(F) = -C_1 - C_2 - C_3 + U_1 = -\$380 + \$1{,}000 = \$620$$

All the information needed to evaluate the exploration project is shown in Figure 7.2. The information is analyzed as follows:

1. Start from the right-hand side, with Stage 3–Observation. If the decision is made to develop the deposit, the expected value of this decision is

$$33.3\% \times (-\$580) + 66.7\% \times \$620 = \$220$$

2. Go one column to the left, Stage 3–Action. To decide whether to develop the deposit after a successful feasibility study, one must compare the expected value of development (\$220) with the expected value of abandonment (–\$80). The optimal decision is to develop the deposit.

3. Go to Stage 2–Observation. There is 40% probability that the feasibility study will give positive results with expected value of \$220, and 60% probability of negative results with cost –\$80. The expected value of completing a feasibility study is

$$40\% \times \$220 + 60\% \times (-\$80) = \$40$$

4. In Stage 2–Action, if exploration is successful, one must choose between completing the feasibility study, with future expected value

$40, or stopping the project, with sunk cost –$10. The optimal decision is to complete the feasibility study.

5. Let us now consider Stage 1–Observation. There is 70% probability of negative results (cost –$10) and 30% probability of positive results (value $40). The expected value of completing Stage 1 is

$$70\% \times (-\$10) + 30\% \times \$40 = \$5$$

In conclusion, the expected NPV of the Stage 1 exploration project is positive, equal to $5 million. The project should be approved at an expected cost of $10 million, with the understanding that the results obtained should be reviewed before going to Stage 2. When Stage 1 is completed, all forward-looking probabilities, costs, and expected revenues should be reestimated.

In practice, it is not likely that Stage 1 will be approved. Analysis of the results summarized in Figure 7.2 shows that the proposed exploration program has some very high-risk components. Even if the feasibility study gives positive results and a mine is being developed, there is one chance in three that the project will fail, resulting in a loss of $580 million. It is unlikely that a reasonable decision maker would go ahead with this project, where the probability of significant loss does not compare favorably with the probability (two chances in three) of a significant gain ($620 million).

This example illustrates the need to define the value of a project on the basis of "utility" as opposed to NPV. The concepts of risk tolerance and project utility are discussed in Chapter 9. Before introducing these concepts, it is appropriate to develop a better understanding of those factors that contribute to risk in project evaluation. This is the subject of the following chapter.

CHAPTER 8

MODELING OF SPACE- AND TIME-RELATED VARIABLES

The input variables whose values are unknown at the time a decision must be made can be classified as either space related or time related. Values that remain consistent in three-dimensional (3-D) space are considered *space related.* Values that are expected to change over time are termed *time related.*

DECISION-MAKING AND SPACE-RELATED VARIABLES

Space-related variables are variables whose value does not change over time, but knowledge of which is incomplete at the time decisions are made. The geologic properties of a mineral deposit do not change over time. What does change over time, and only as a result of getting additional information through exploration and data analysis, is our understanding of these properties. The geologic properties that are of interest are likely to include one or more of the following: survey information, structures, petrology, mineralogy, alterations, geochemistry, percentage of minerals of potential economic value, percentage of co-products and contaminants, specific gravity, geometallurgical properties, and so forth. Following are several reasons why information about deposit geology is limited at the time decisions are made:

- The need for specific information may be considered irrelevant. However, the decision as to whether additional information is relevant must be made with care. What is irrelevant to a geologist or a mining engineer may be highly significant to a metallurgist or an environmental engineer. Traces of contaminants may be considered too small to require systematic testing. But different parts of the deposit are likely to have different percentages of deleterious elements, and a flotation plant may concentrate deleterious elements to levels that exceed those specified in sales contracts.

- The competence and level of experience of the team responsible for project evaluation may not be appropriate, resulting in not recognizing the relevance of significant risk variables.
- The cost of obtaining additional information is high. Geological properties vary throughout the deposit. Information about the relevant geological parameters can be obtained by drilling, sampling, logging, assaying, and testing. But this process is costly and time-consuming. At some point, the decision maker is expected to decide that the value of additional information does not justify the cost.
- The quality (precision and accuracy) of available information is limited because errors are bound to occur during surveying, sampling, sample preparation, assaying, laboratory analysis, sample logging, and data reporting.
- It is not sufficient to have reliable data. Analysis of this data, including statistical analysis and development of models that represent the properties of the deposit in three dimensions, is necessary. Most decisions are based on models from which production schedules are developed as needed to estimate expected cash flows. Models are imperfect.
- Computerized systems are used to model deposits. Consulting companies sell the necessary software and train users accordingly. Often, complex geostatistical algorithms are incorporated in these systems. There are always underlying assumptions that must be satisfied for these algorithms to be valid, the most common one being the assumption of stationarity. Deposit modelers are not always aware of these assumptions or of how important it may or may not be that they are contextually validated.

- Geostatistical methods are available to quantify the uncertainty with which the 3-D properties of a deposit are known. These methods include deposit simulation, which consists of creating different realizations of what the deposit properties might be between data points. These simulations take into account known information, such as the location and value of available samples, geological zoning, and the statistical properties of the sample values (mainly frequency distribution and variogram) within these domains. There are limitations to the capabilities of a computer program to simulate a deposit, some resulting from the underlying assumptions under which the software algorithms are valid and others being a reflection of the capabilities of those who use the software.
- Software developers are continuously improving their product and are highly competent in making the use of complex algorithms easily available to a wider range of geostatistical experts. However, ease of use does not remove the need for experts to understand the conditions under which use of these algorithms remains valid. Being able to use computerized systems that include complex geostatistical algorithms does not equate to the ability to generate meaningful deposit models.

One field of geostatistics that is becoming increasingly popular is deposit simulation (COSMO n.d.; CCG 2014), including simulation of geology, grade distribution, geometallurgical properties, structures, and so forth. Indeed, use of simulation for risk analysis and decision-making can be highly effective. However, the validity of deposit modeling methods, whether deterministic or probabilistic, depends on specific stationarity conditions being satisfied. One underlying assumption is that the statistical properties of the variable of interest are reasonably constant throughout the area being evaluated. Therefore, deposits are divided into domains. Domain definition is an art, which has significant consequences. Small domains result in meaningless estimates. Large domains may not internally satisfy minimum stationarity conditions. Different variables, such as the grade of different minerals within the same deposit, may be the subject of different geologic controls and may require definition of different domains.

Stationarity conditions can become increasingly demanding when the more advanced geostatistical methods are used. For example, a simulation process may consist of

- Transforming the known frequency distribution of the available sample values, say, x = Cu%, into a normally distributed variable, y;
- Analyzing the statistical properties of the normalized values y;
- Simulating what the value of the normalized variable might be between sample points; and
- Transforming the resulting simulated values of y back into the original x = Cu% space.

Underlying assumptions include that the same transformation applies to the entire area being modeled and that the variogram of the normalized values is the same throughout. Provided sufficient data is available, statistical methods can be used to test the validity of these assumptions. It is not uncommon for deposit modelers to ignore or not be aware of the underlying assumptions and to assume that using more advanced geostatistical methods on relatively small data sets will result in better deposit models. Consultants encounter situations where clients will require that advanced geostatistical methods be used, not because their use is appropriate, but because the results obtained can be "sold" to management as being of superior quality. Such practices should be condemned.

Variograms are used to quantify the relationship between sample values when the distance between samples increases. Because of lack of close-spaced samples, variograms are often poorly defined for short distances. Such distances are those that are most important to obtain meaningful simulations, especially when simulation is used to assess how mine selectivity will influence tonnage and grade of material above cutoff grade. Instead of assuming that the variogram is perfectly known, different realizations of the grade distribution should be simulated using a range of possible variogram models.

When modeling space-related variables, one should keep in mind the strengths and weaknesses of the different modeling methods that can be used. A main feature of traditional methods, such as ordinary kriging, is that they produce conditionally unbiased estimates: If we mined all the blocks estimated at 0.5% Cu, the average grade of these blocks will be 0.5% Cu. Some blocks will be higher grades, some lower, but the average will be 0.5% Cu. This property is great if we know we are going to mine all these blocks, as may be the case when estimation is made from blastholes to decide which material should be sent to the processing plant.

However, this assumption is not valid when modeling a deposit from exploration drill holes. Additional information will be obtained before the decision to mine the blocks is made. When such information is available, some blocks (or some parts of each block) will be treated as waste, which were originally estimated above cutoff grade. Conversely, some blocks previously estimated as waste will be mined and processed. The tonnage and grade of material that will ultimately be mined and processed will depend on the amount of information available at the time mining takes place and is likely to be significantly different from that estimated by ordinary kriging using widely spaced exploration drill holes. This phenomenon is known as the information effect. The more recent geostatistical methods are becoming increasingly effective at resolving such weaknesses inherent to historical methods.

Additional challenges are encountered when modeling mineral deposits, which differ depending on whether mining will be by open-pit or underground method. In open-pit mines, all the material within the pit limits will be extracted. To value the pit, it may be sufficient to model the proportion of reasonably large areas, defined as "panels," that will be minable above cutoff grade, taking into account the applicable ore control method. Statistical methods have been developed accordingly. The exact location where the high-grade material is located within each panel will be determined when mining takes place.

In underground mines, the situation is different. Applicable modeling methods depend on the mining method to be used[1]. The cost of getting information is higher. Complexity increases with the need for selectivity. Stopes that are designed from surface drill holes may differ significantly in location, shape, and grade from those that will be delineated and mined once underground access is obtained. However, the deposit model that will be used to

1 The modeling method will not be the same if the deposit is to be mined by cut-and-fill or block caving.

determine the economic value of the project must be such that the location, size, and grade of the stopes that will be mined are determined with sufficient accuracy. Mining method, mine infrastructure, achievable mining rate, size of processing plant, expected head grade, and cash flow are all determined from this information. How to model deposits for underground mining can also be considered a statistical problem, but with constraints that are different from, and often more demanding than, those faced when modeling deposits that are open-pit minable.

DECISION-MAKING AND TIME-RELATED VARIABLES

Properties of Time-Related Variables

Another major source of uncertainty in mining results from changes that will occur over time during the project life. Typically the value of a time-related variable is known at the time a decision is made, but this value is expected to be different one month, one year, or five years later when mining operations take place. Both the magnitude and timing of such changes are usually not known.

Variables that fall in this category include the value of the product to be sold (gold or copper price, price of coal or iron ore, etc.) and the cost of input (labor, consumables, and energy). Other time-related variables are of environmental (earthquakes, floods, drought, etc.) or social (strike, nationalization, social unrest) nature. The future value of some risk variables can be controlled using such means as contracts, forward sales, or partnerships. Other variables such as metal spot prices cannot be controlled.

The main question that must be answered when dealing with risk caused by unforeseen changes is whether the project has enough built-in flexibility to adapt to such changes. This is the same question that has to be answered when dealing with space-related uncertainty—except that one cannot sample the future. Nevertheless, it is possible to simulate a number of equally probable realizations of the future values of the time-related variables and calculate the project utility for each realization. The risk associated with the time-related variables is determined by analyzing the probability distribution of these simulated utilities.

The simplest assumption is that the project has no flexibility, and that the original design, including mine plan, processing capacity, productivity, costs, volume of sales, and so on, will not or cannot be changed even if the value of the risk variables changes from that originally assumed. In practice, flexibility is nearly always available and changes can be made, some on a short-term basis, others with longer-term consequences, which make it possible to minimize the negative impact or maximize the positive impact of changes in the value of input variables. Assuming lack of flexibility in the operating conditions can be used to define an upper limit to the risk presented by the project as designed. An optimal project is one that has enough built-in flexibility that it can adapt to likely changes in the value of input variables and take advantage of opportunities resulting from these changes.

Time-related variables can behave in different ways:

- Their value is constant over time. These variables do not present any risk, yet few variables fall into this category. In stable legal environments, royalties, long-term contracts, and tax rates may fall into this category.
- Their value is constant over a given time period, such as the duration of a contract, but are expected to change either up or down at the end of the period. The probability distribution of the expected changes must be modeled.
- Their value changes in an apparent random fashion up and down without showing any systematic trend. This random variability must be modeled with particular attention given to possible extreme values.
- Their value is expected to show systematic increases or decreases over time, with random noise around these trends. The main concern should be with timing, duration, and magnitude of these trends, all of which can only be modeled probabilistically.
- Their value has a high probability of remaining stable over long time periods but may change abruptly and without warning at any given time. Earthquakes, volcanic activities, and terrorist actions fall into this category.

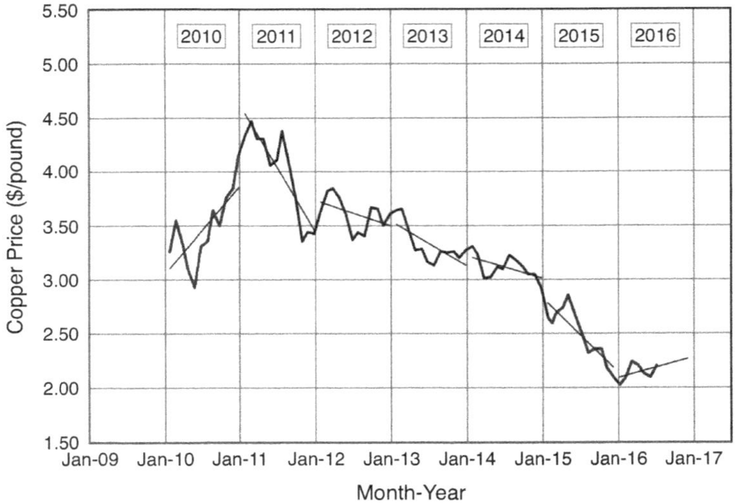

FIGURE 8.1 Monthly copper prices, 2010–2016

Metal Prices as Risk Variables

The variability in metal prices can be interpreted as composed of periods with relatively stable prices followed by significant periods of increasing or decreasing prices, leading to a new plateau, either lower or higher. Long-term trends in copper and gold prices are illustrated in Figure 8.1 and Figure 8.2, respectively. Copper prices per pound increased rapidly from a $3.00-to-$3.50 range in 2010 and to a $4.10-to-$4.50 range in 2011. The next plateau lasted 18 months when prices varied between $3.40 and $3.80. Prices fell again in 2013, 2014, and 2015. Even when copper prices were relatively constant, such as from the end of 2011 to early 2013, the monthly variations remained significant.

Figure 8.2 shows that gold prices remained essentially constant, just below $400 per ounce over a very long 15-year time period before decreasing to less than $300 in 2000 and then skyrocketing to $1,700 in 2012. This sharp increase was followed by just as sharp a decrease to $1,100 in 2015. Shorter-term trends in gold prices are discussed later.

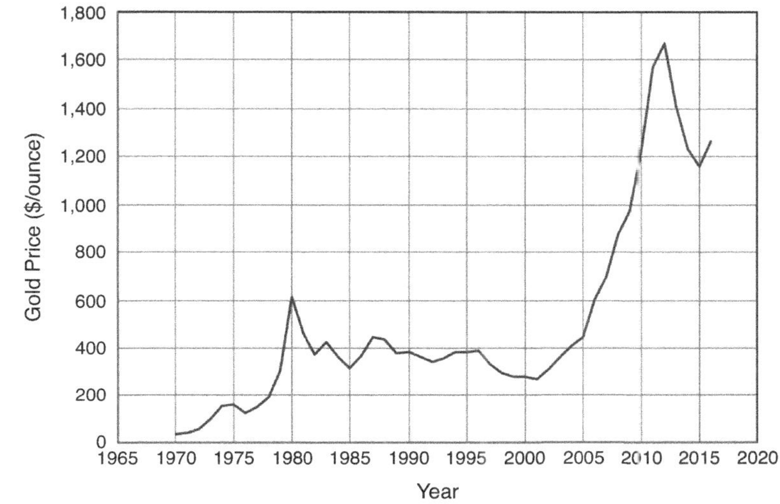

FIGURE 8.2 Yearly gold prices, 1970–2016

TRENDS AND RANDOM NOISE IN TIME-RELATED VARIABLES

There are many ways to classify the various sources of uncertainty. A distinction has already been made between space-related and time-related uncertainty. Another classification that is worth mentioning is based on the difference between what could be considered as *predictable uncertainty* (more commonly referred to as *trends*) and *unpredictable uncertainty* (or *random noise*). To explain what is meant by these terms, one can consider Figure 8.3, which shows daily variations in gold prices from January 2010 to August 2016.

During 2010 and 2011, the price of gold was increasing steadily and one could have projected gold prices upward from one month to the next, assuming that historical trends will continue over time. Indeed, in 2011 many analysts, investors, and mining companies were predicting that the gold price would reach $2,000 per ounce in 2012. Such a prediction would have been seriously flawed, because some fundamental changes occurred in the second half of 2011 that were not anticipated by most.

A period of high gold price volatility started in the second half of 2011, which lasted until the end of 2012. During this period, one could have projected prices to be highly variable while remaining between $1,600 and $1,800 per ounce. But once again, a major shift occurred in 2013 when prices fell from $1,700 to $1,200. If this trend had been projected downward, the gold price would have been expected to reach $1,000 by mid-2014. But instead, a new equilibrium was reached, with prices varying in a relatively narrow range, mostly between $1,250 and $1,350. Unfortunately, gold prices fell to a still lower level in late 2015, falling below $1,200 per ounce. This downward trend reversed itself in 2016.

One could consider five periods with reasonably predictable statistical properties: 2009–2011, 2011–2012, 2013, 2014–2015, and 2016. Within each period, trends and noise above and below trends could have been reasonably predicted by projecting observations from one year to the next. The main challenge would have been to predict the sharp changes that occurred when moving from one period to another. It is not uncommon, indeed arguably quite usual, for market analysts to overpredict future prices when recent trends are upward and underpredict them when trends are downward.

Risk takers might assume that the rapid increase in gold price observed from January to September 2016 will continue through 2016 and beyond. Such an assumption would be similar to those made in 2010 and 2011 and could have the same negative consequences when the trend reverses itself. When estimating a project risk, a wise manager (and a wise investor) would study the risk to the project not only if trends continue in their current direction but also if unpredictable disruptive changes occur randomly during the life of the project. Disruptive changes are changes that radically change the business environment and can only be accommodated by taking new strategic directions. Failure to adapt to disruptive changes, such as those that occurred in the gold price during the period 2012–2013, can result in significant opportunity losses.

Predicting future metal prices or changes in the cost of fuel or labor is both an art and a science. Consulting companies specialize in supplying their view of what prices will be, with moderate success. Statistical methods, such as time series analyses and geostatistics (Appendix B), can be used to analyze past behavior in order to predict future behavior. However, such methods do not

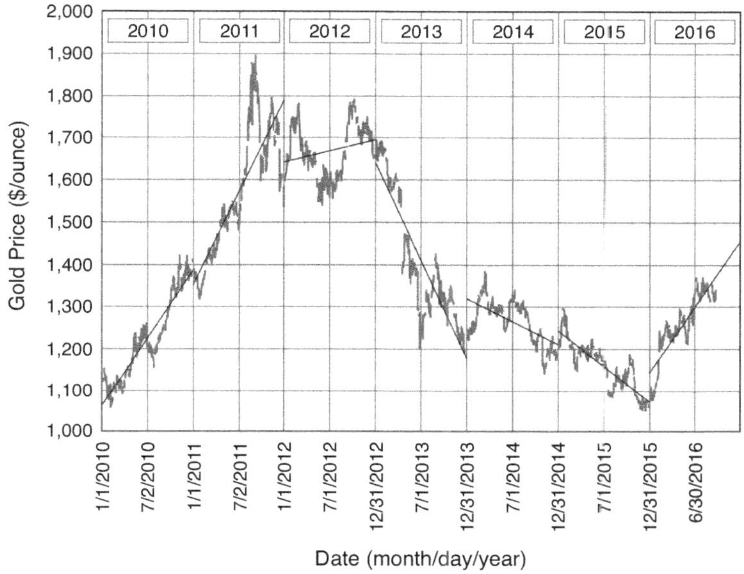

FIGURE 8.3 Daily gold price in U.S. dollars per ounce, 2010–2016

take market fundamentals into consideration. Expert advice based on market studies and statistical analyses offer some of the best, albeit not necessarily highly reliable, estimates of the future. Advice from a number of experts with occasionally conflicting opinions may be the better way to define possible future scenarios.

Some risk variables, such as the occurrence and magnitude of earthquakes, floods, catastrophic fires, landslides, and so on, are unpredictable. Different statistical methods are applicable depending on the nature of the variable whose probability of occurrence is to be modeled. Poisson distributions are useful to represent the probability of occurrence of unpredictable events.

The impact of climate change during the life of a project must be taken into account. What may have been historically classified as a 100-year event (an event that has a 1% chance of occurring in any given year) may become a 50-year event. In arctic regions, mines that depend on winter ice roads may find that access is limited to shorter periods as a result of warmer weather. Similarly, river access may be significantly disrupted by longer than normal draughts or frequent floods. The work of engineers in properly evaluating the probable effect of climate change on a project may be compromised when political beliefs interfere with scientific studies.

MITIGATING THE IMPACT OF RISK VARIABLES

The following steps must be taken to mitigate the impact of risk variables:

1. Determine which are the input variables that could significantly (positively or negatively) impact the value of a project. This assessment is best completed using sensitivity analyses combined with experienced project reviews.
2. Estimate the values (range or values, as when considering ore grade; or specific events such as catastrophic weather events or nationalization) that these variables might take.
3. Estimate the probability, objective or subjective, that these values will be realized.
4. Determine which method, if any, can be used to change the impact of the possible events on the project value, or the probability that these events will occur.
5. Decide whether risk mitigation is sufficient to justify the cost of mitigation.

Two methodologies can be combined to evaluate and mitigate risk: simulation and business strategy.

Risk Assessment and Simulation

The steps that can be taken to quantify and reduce the risk resulting from incomplete knowledge of the value of space- and time-related variables can be summarized as follows (see Figure 8.4):

1. Define a base case. Make initial deterministic assumptions concerning the most likely value these variables will take when the deposit is mined.
2. Design mining and processing facilities and estimate production schedule, capital and operating costs, and expected life-of-mine cash

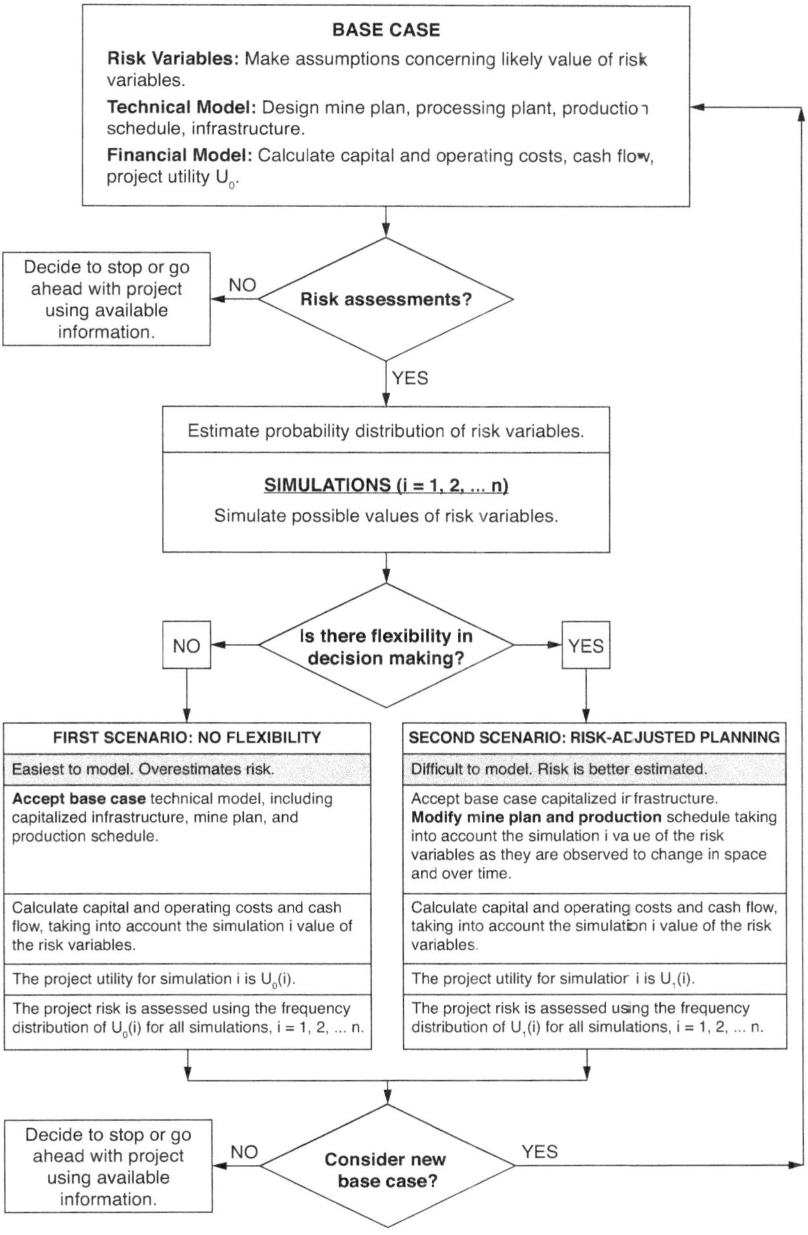

FIGURE 8.4 Risk assessment with or without risk mitigation

flow. Key indicators, such as net present value (NPV), internal rate of return (IRR), and payback period, are used to determine the utility U_0 of this base case.[2] If risk is ignored, the project evaluation ends here.

3. Simulate possible values of the time- and space-related variables. These values differ from those used in the base case. One may consider n simulations, i = 1...n.

4. For each simulation i, consider two scenarios. The first scenario consists of assuming that the base-case mine plan and production schedule remain unchanged over time and are independent of the values of the risk variables. The utility of this scenario, which assumes no flexibility, is $U_0(i)$.

5. Analyze the frequency distribution of the n realization $U_0(i)$, i = 1,n. This distribution defines the risk associated with the project as designed, assuming that no adjustments are made over time when conditions change.

6. The second scenario consists of developing a risk-adjusted production schedule, which takes into consideration actions that could be taken to minimize the negative impact and maximize the positive impact of the input variables as simulated. This scenario assumes that at the beginning of each year, the decision maker knows the past and current value of the input variables but does not know their future value. The utility of this risk-adjusted scenario corresponding to the i-th simulation is $U_1(i)$, which should be higher than $U_0(i)$.

7. Analyze the frequency distribution of the n realizations $U_1(i)$. From this distribution, one can determine the risk-adjusted utility of the project, including the probability of project failure and of the project performing above expectation. This distribution is dependent on the base-case assumptions around which all simulated variations were modeled.

8. To optimize the project design, one must consider different base-case assumptions. For example, one may consider a new base case that requires less initial capital but higher operating costs, or a base

2 The factors that define the utility of a project are discussed in Chapters 9 and 10.

case with higher capital expenditures and lower operating costs. Each base case is then analyzed using the same approach as just described. The various base cases are then compared, and the one that has the highest risk-adjusted utility should be retained.

This approach can be extremely time-consuming unless only a few base-case scenarios are considered that are tested against a small number of simulations. Considering only the most extreme cases, such as those which give the highest and lowest metal prices that one could expect during the life of the project (often respectively called the optimistic or Blue Sky and pessimistic scenarios), can significantly reduce the number of analyses but may lead to incorrect decisions. It is worth mentioning that universities and software companies are working diligently to develop theoretical and practical ways to complete the optimization process in a timely fashion, making it possible to consider a larger number of simulations simultaneously. Research is mostly taking place in the optimization of pit design when spatial uncertainty related to the deposit properties is taken into account, or when future costs and prices are likely to change in a random fashion.

Risk Mitigation and Business Strategy

A company can mitigate project risk in many ways, which depend on the company objectives and the type of variables whose imperfect knowledge may result in a risky situation. Many examples are given throughout this book. Strategic decisions that are worth mentioning here include the following:

- Base decisions only on NPV, IRR, payback period, and other financial indicators. This is the simplest and most commonly used approach. Even when other parameters are considered, financial characteristics always influence the decision-making process.

- Take into account the operating costs. A company might consider acquiring properties whose operating costs are in the lower half of the industry cost curve, even if the capital cost is high. This approach makes it likely that the property will continue to generate a positive cash flow when markets are depressed.

- Bring in a partner to develop a project that requires more capital than the mining company is willing to invest. This approach takes into account project size, capital availability, borrowing power, other projects in the company's portfolio, and the need for diversification to mitigate risk.
- Bring in the government as an equity partner rather than a royalty holder. This approach more closely relates costs to market conditions. Whether the government owns royalties or equity, the political risk remains that the government might want to renegotiate its share of ownership when project returns increase beyond original expectations.
- NPV optimization studies often show that the theoretically best solution consists of high production rate and high initial capital investment, with low operating costs, higher reserves, and increased cash flow. However, risk management might show that a phased investment is preferable, to reduce current capital requirements and at-risk initial capital, and in some cases bring forward production and positive cash flow. The risk taken by phasing successive increases in production might be more manageable than a one-time high capital investment.
- Many companies have one or more key assets whose loss would be devastating from a financial, social, or "public image" point of view. The risk that can be taken when making decisions related to such properties is likely to be significantly less than for other less strategically important properties.
- Using international sources of funding can be an effective way to increase local and national credibility and reduce the risk of creeping or outright nationalization. The International Finance Corporation can be considered as a partner for large-scale sustainable investments, providing financing and advice for private sector clients, and helping governments adopt effective regulations and strengthen their capacity to manage industries across the value chain.

TABLE 8.1 Use of simulation to evaluate risk associated with time-related variables

	Year 0	Year 1	Year 2	Year 3	Year 4	Year 5
	Base Case (Feasibility Study)					
Base-case gold price ($/ounce)	**$1,000**	**$1,000**	**$1,000**	**$1,000**	**$1,000**	**$1,000**
Cash flow ($ millions)	($400)	$180	$275	$100	($40)	
NPV @ 9%	**$45**					
	Simulation 1					
Simulated gold price ($/ounce)	**$970**	**$1,205**	**$1,310**	**$1,430**	**$1,400**	**$1,260**
Unadjusted cash flow ($ millions)	($400)	$242	$430	$315	($40)	
NPV @ 9%	**$398**					
Risk-adjusted cash flow ($ millions)	($400)	$246	$437	$314	$260	($60)
NPV @ 9%	**$581**					
	Simulation 2					
Simulated gold price ($/ounce)	**$940**	**$970**	**$900**	**$790**	**$850**	**$900**
Unadjusted cash flow ($ millions)	($400)	$171	$225	($5)	($40)	
NPV @ 9%	**($86)**					
Risk-adjusted cash flow ($ millions)	($380)	$180	$254	$35	($30)	
NPV @ 9%	**$4**					

ASSESSING PROJECT FLEXIBILITY: AN EXAMPLE

The risk assessment and mitigation process is best illustrated using an example such as that summarized in Table 8.1. In this situation, a gold deposit is considered for development, and the future gold price is unknown. A base-case production schedule is developed assuming a constant gold price of $1,000 per troy ounce. In this base case, the duration of the project construction period is one year (Year 0). This is followed by three years of revenue-generating operations (Years 1 to 3) and then one year for mine closure (Year 4). The expected cash flow is estimated using the assumed constant gold price. The project's NPV, calculated using a 9% discount rate, is U_0 = $45 million. This base case is summarized at the top of Table 8.1.

An alternative gold price scenario is considered, defined as Simulation 1 ($i = 1$) in Table 8.1. In most years, the gold price is significantly higher than that assumed in the base case. The assumption is first made that the base-case investment and production schedule remain unchanged. Because the gold price is higher than expected, the cash flow is significantly improved. The corresponding NPV is $U_0(1)$ = \$398 million.

Using the same Simulation 1 gold price scenario, the assumption is made that the initial capital investment remains the same, but mine management adjusts the production schedule yearly to benefit from that year's actual gold price. At the beginning of Year 1, mine management looks at the prevailing gold price (\$1,205) and makes adjustments to the production schedule to take advantage of this price. When such adjustments are made, the manager does not know what the gold price will be in Year 2 and beyond but can make estimates of what these prices might be, based on recent history. Such adjustments may include a lower cutoff grade, stockpiling of low-grade material, and the design of a larger pit, thus increasing mineral reserves and life of mine. Consideration could be given to increasing mine and processing plant capacities, allowing a higher production rate and resulting in lower unit operating costs. The ultimate result is an increase in profitability, which in Year 1 goes from \$242 million to \$246 million. Other improvements are made at the beginning of Years 2 and 3, and, as a result of good husbanding of resources, production lasts one more year and closure takes place in Year 5. The larger operation imposes higher closing costs: \$60 million instead of \$40 million. The NPV of this risk-adjusted production schedule is $U_1(1)$ = \$581 million.

Another simulation is also summarized in Table 8.1. In Simulation 2, the actual gold price is systematically lower than the \$1,000 per ounce assumed during the feasibility study. If no adjustment is made to the production schedule, the project's NPV will be negative: $U_0(2)$ = –\$86 million. This does not mean that the project should be stopped, given that the decision to build the mine was made before realizing that the price was going to be too low to justify the investment, and yearly cash flows are positive. As a minimum, if capital and operating costs remain the same, the mine should not operate in Year 3, when operation would result in a \$5 million loss. In practice, management will do what it can to reduce capital and operating costs and maintain profitability under the new economic conditions. This may include

- Reducing the capital cost from \$400 million to \$380 million, which may be done by eliminating or postponing all non-essential capital expenditures;

- Decreasing operating costs by reducing staff, postponing maintenance, and eliminating all nonessential expenditures;
- Redesigning the pit by eliminating the more expensive pushbacks;
- Increasing the cutoff grade; and/or
- Lowering the closure cost as a result of a smaller pit and reduced environmental impact.

The risk-adjusted NPV is $U_1(2)$ = \$4 million.

CHAPTER 9

RISK TOLERANCE AND UTILITY FUNCTION

Understandably, most people exhibit some aversion to loss, and the same can be said about managers and companies. In Chapter 7 it was shown that optimization of the expected net present value (NPV) of a project is not necessarily a reliable criterion for decision-making, because expected values give the same weight to a loss and a gain of equal magnitude. This chapter investigates how risk tolerance influences the decision-making process and why there is a need to define the "utility" of a project as a better measure of value than criteria of a purely financial nature.

DEFINITION OF RISK TOLERANCE

Consider a situation where a mining company owns a mineral property and has to decide between two options:

- **Option 1:** There is a willing buyer who is offering to pay \$50 million to acquire all rights to the property as well as all related liabilities. Accepting this option presents no risk.
- **Option 2:** The mining company could keep the property and continue developing it. Information available to date indicates that there is equal probability that the return on investment will be either \$0.0 (breakeven situation) or \$100 million.

The expected value of the second option is $0.5 \times 0.0 + 0.5 \times 100 = \$50$ million, equal to the guaranteed return of selling the property (Option 1). A company that is insensitive to risk, or risk neutral, would consider both options as equivalent. A company that is risk averse would give preference to the guaranteed $50 million payoff rather than accepting a 50% probability of a zero return. Depending on the level of risk aversion, the company would even accept a guaranteed payment of less than $50 million rather than accepting the risk of no return. Conversely, a risk-loving company would accept Option 2 even if the guaranteed payment was more than $50 million (but less than $100 million).

Notably in this example, one assumes that the expected dollar return is the only measure used by the mining company to choose between options. If other objectives were taken into account, such as maintaining or increasing production or market share, a risk-averse company might take the risky Option 2 even if the expected dollar value of this option was less than the guaranteed $50 million. This should not be interpreted as an indication that the company is risk loving. When taking into account the nonmonetary benefits related to developing the property, the value of Option 2 is higher than that of Option 1. When comparing options, one must take into account all costs and benefits related to each choice. Expected dollar value is only one of a number of measures that define the utility of a project.

DEFINITION OF UTILITY FUNCTION

The expected dollar value of a project is not a perfect representation of what the project is worth to the company. Expected dollar values do not take into account the company's risk tolerance. Everything else being equal, two companies that consider the same project will value it differently, depending on the extent to which they are risk averse, risk neutral, or risk loving. For these reasons, decision-making theorists have introduced some 50 years ago the concept of the "utility function." In its simplest form, the utility function is used to transform the dollar value x of a project, usually its NPV, to a number U(x) that takes into account the risk tolerance of the company. This number is defined as the utility of the project, and the corresponding unit of measurement is the util/utils.

Utility functions are expected to increase with the dollar value. A project of zero dollar value is usually given a utility U(0) equal to zero. The scale used to define the utility function is arbitrary. To be meaningful, a utility function must have additional properties that are best explained by examples such as the following.

A mining company is considering acquisition of a mineral property. Exploration completed to date has discovered mineralization, which has not yet been fully delineated. Company experts estimate that, depending on what additional drilling would prove, the project value could be either \$100 million or \$500 million with equal probability. The expected value of the property is 50% × 100 + 50% × 500 = \$300 million. Would the company pay the same amount for this property as it would for a safer investment that is guaranteed to return \$300 million? Probably not. The mineral property is worth less than the safe investment, but how much less?

Expected NPV does not take into account risk tolerance. In the preceding example, the average utility of two equally likely gains, \$100 million or \$500 million, is less than the utility of a sure gain of \$300 million. Consider Figure 9.1. In this figure, it is assumed that the utility of \$500 million was arbitrarily set at 100 utils, and that of \$100 million at 20 utils. These utilities apply to projects with guaranteed returns of \$500 and \$100 million, respectively. The expected utility of a project that has equal probability of returning \$100 million or \$500 million is

$$U_1 = 50\%\ U(\$500) + 50\%\ U(\$100) = 50\% \times 100 + 50\% \times 20 = 60 \text{ utils}$$

If a guaranteed return of \$300 million is worth more, the utility of such a return must be higher than $U_1 = 60$ utils:

$$U_2 = U(\$300) > U_1$$

The implication is that a utility curve is a concave curve, as shown in Figure 9.1. On this curve it is assumed that U(\$300) is 67 utils.

The downward curvature of the utility function is likely to increase when the NPV decreases, and more so when the NPV is negative. This is shown in Figure 9.2. The reason is that risk aversion increases when the value of a project decreases, and that aversion to losses (negative NPV) increases when the magnitude of potential losses increases.

If a decision maker is risk neutral, the expected NPV is considered a reasonable number to define the utility of a project, and the utility function is a straight line. At the other end of the risk-tolerance spectrum, the utility function of a risk-seeking decision maker would curve upward; the risk associated with low returns would be given little weight, whereas high returns would be given exaggerated weights.

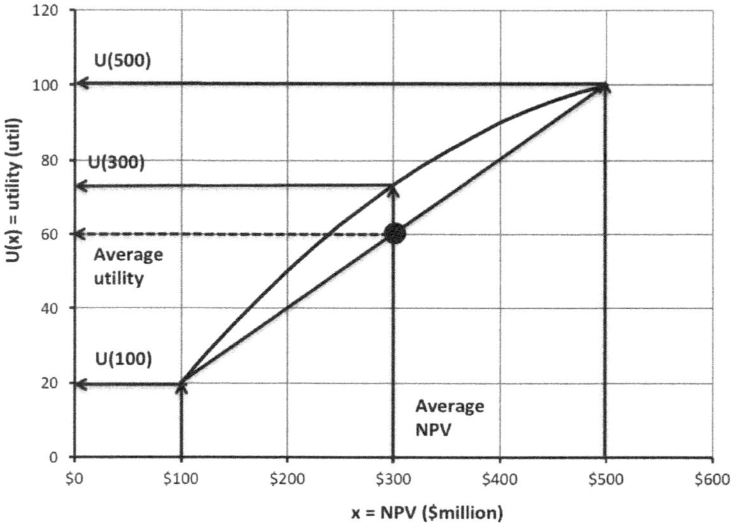

FIGURE 9.1 In risk-averse situations, the average utility is less than the utility of the average NPV

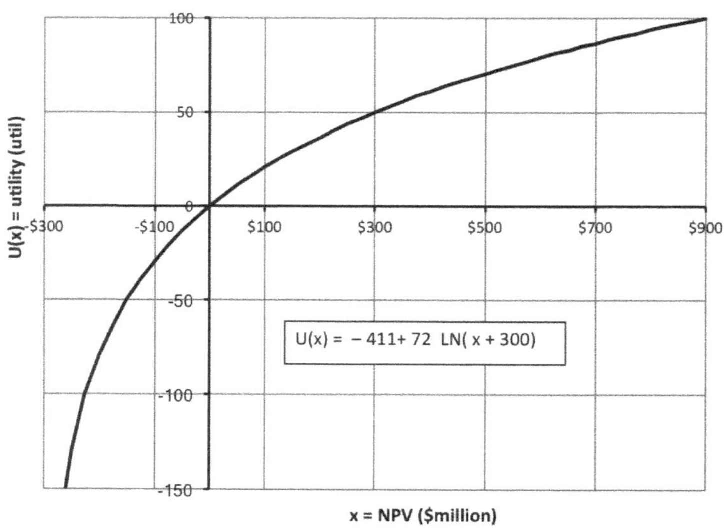

FIGURE 9.2 Possible shape of utility function reflecting increasing aversion to downward risk

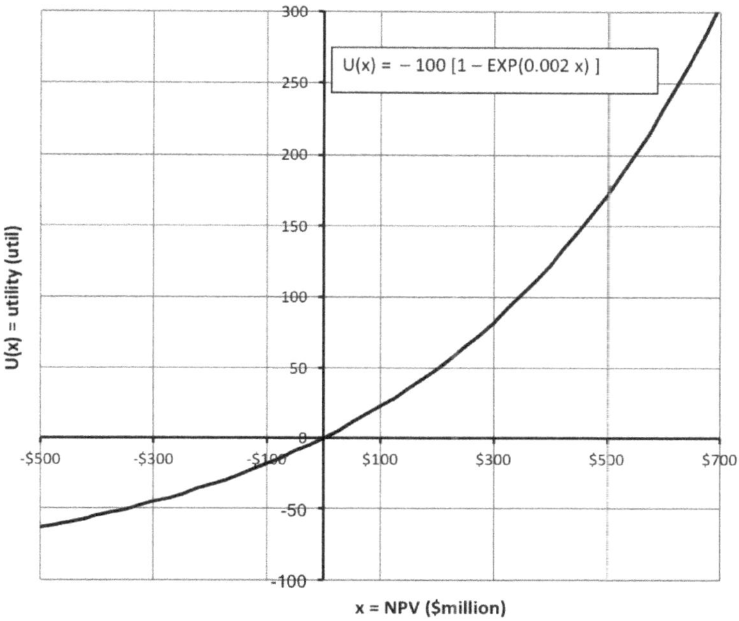

FIGURE 9.3 Possible shape of utility function for a risk-loving decision maker

For example, consider a risk-loving decision maker whose utility function is as shown in Figure 9.3. The utility of a loss of \$500 million is U(–\$500) = –63 utils, while that of a gain of \$500 million is U(+\$500) = +172 utils. This decision maker would go ahead with a project that has two chances in three to result in a loss of \$500 million and only one chance in three to show a gain of \$500 million. To him or her, the utility U_1 of going ahead is positive:

$$U_1 = \tfrac{2}{3} \times (-63) + \tfrac{1}{3} \times (+172) = 15 \text{ utils}$$

Las Vegas gamblers are good examples of risk-seeking individuals. A gambler might consider that, even if the probability of a high return is very low, spending a significant amount of money for a chance to win is justified. The gambler may consider spending \$1,000 for a one-in-a-million probability of gaining \$1,000,000. The expected dollar value of playing the game is a loss of \$999, calculated as follows.

- Expected dollar loss of playing the game:

$$10^{-6} \times \$1{,}000{,}000 - (1-10^{-6}) \times \$1{,}000 = -\$999$$

The Las Vegas gambler considers that the expected utility U_1 of playing the game is positive, even if the utility U_2 of a guaranteed \$999 dollar loss is negative.

- Expected utility of playing the game:

$$U_1 = 10^{-6} \times U(\$1{,}000{,}000) + (1-10^{-6}) \times U(-\$1{,}000) > 0$$

- Expected utility of a guaranteed \$999 loss:

$$U_2 = U(-\$999) < 0$$

These conditions can only be satisfied if the gambler's utility function curves upward. This is not likely to be the case if the gambler earns \$25,000 a year with no significant net worth. On the other hand, a multimillionaire may play the game for his or her own amusement or to impress observing friends.

IS MINING A RISK-SEEKING OR RISK-AVERSE BUSINESS?

Some consider that mining is, or should be, a risk-seeking business. Mining professionals, investors, and financial analysts will occasionally state that the management of one or another mining company is not taking enough risk and is rejecting projects without looking at the possible upside. But are such criticisms justified if the likelihood of significant upside is undefined or extremely low while the cost of acquiring or developing the project is assured and high? Conversely, when a project fails, management may be criticized as risk-loving, having taken too much risk. Some business analysts may criticize the same managers as being risk averse in some situations and risk loving in others. Such criticism is more likely to reflect limited understanding of the different risk associated with each project rather than irrational behavior on the part of the managers.

Senior management needs to be conscious of these criticisms, whether justified or not. Decisions should not appear to be made arbitrarily, based on feelings and emotions. Sound management requires understanding that risk can be managed, provided it is well informed. As shown in this chapter and throughout this book, making sound decisions is a rational process designed to control risk and manage an investor's monies in a responsible way.

A company will usually have a final tolerance to the magnitude of the losses that it is willing to accept. A project that may fail and whose cost of failure would exceed that which can be tolerated should be rejected. This is reflected in Figure 9.2 where the utility of a project becomes infinitely negative when its estimated NPV approaches a loss of $300 million. If decisions are made on the basis of utility rather than NPV, any project that could result in such a loss will be rejected no matter how small the probability of failure might be.

The magnitude of losses that are acceptable to project managers might be somewhat lower than that acceptable to the company as a whole. A loss that, if incurred, would result in the manager losing his or her elevated position of responsibility, and the personal rewards that come with it, would be of great concern to this manager even if such a loss would have few corporate repercussions. On the other hand, a project manager may perceive that he or she will be immediately rewarded for bringing a project to the next stage. The manager may give higher value to assured, personal, and immediate rewards than to the longer-term, uncertain consequences of project development. In such situations, a manager's values would conflict with those of the company and its shareholders. The company should avoid creating an environment where such situations may occur. The rest of this discussion assumes that the management's values are the same as those of the company as a whole.

MATHEMATICAL REPRESENTATION OF UTILITY FUNCTIONS

Determining the shape of the utility function that applies to a specific company or individual is not a simple matter. There is no reason why the same mathematical equation could be used to represent the utility function of all individuals and all companies. A variety of approaches to the determination and modeling of utility functions can be found in the literature. Most mathematical models, including those used in this chapter, assume that the utility of a project is a function of a single variable such as the estimated NPV. This assumption is somewhat simplistic. As can be seen later in Chapter 10, the utility of a project depends on a number of criteria, including financial, social, and environmental, some of which cannot be expressed in dollar terms.

The most significant contribution of the concept of utility is the understanding it brings to the manager on how best to make sound decisions under uncertainty:

- The utility of a project is a function of the risk tolerance of the decision maker.
- Financial measures and corresponding uncertainty are necessary but not sufficient to define the utility function.
- The utility of a project decreases faster than its financial benefits.
- There is likely to be a limit to the maximum loss that the company can accept. Any project that may result in such a loss should be rejected even if the probability is very low.
- For rational decisions to be made, companies should ensure that the company utility function—including values and objectives as defined by the board of directors, the senior management team, and those of its project managers—are understood and identical.

A mathematical equation that can be used to represent the utility of a project U(x) as a function of its NPV x is as follows:

$$\text{If } x = 0 \quad U(x) = 0$$

$$\text{If } x > x_{min} \quad U(x) = a + b\,\mathrm{Ln}(x - x_{min})$$

$$\text{If } x < x_{min} \quad U(x) = U_{min}$$

x_{min} = minimum tolerable NPV (negative number measured in dollar terms)

U_{min} = utility of projects with NPV less than or equal to x_{min} (negative number measured in utils)

a = coefficient chosen so that U(x) = 0 when x = 0

b = scale factor to convert dollar values into utils

If U_{min} is chosen small enough, it will guarantee that no project will be accepted that has even a very small probability of resulting in an NPV loss approaching x_{min}. The value of the scale factor b is defined by specifying which arbitrary value U_0 the utility function $U(x)$ should be given for a specified NPV value $x = x_0$. The value x_0 should be positive and sufficiently high to obtain a utility curve that is meaningful within the range of possible project NPVs. The value given to U_0 defines the utility scale and can appropriately be set at 100 utils.

x_0 = positive project NPV value, chosen to obtain a meaningful utility function within the range of possible NPVs

U_0 = $U(x_0)$, arbitrarily defined utility of project for x equal to x_0

The values used in Figure 9.2 and throughout the rest of this chapter are defined as follows:

$$x_{min} = -\$300 \text{ million}$$

$$U_{min} = -1{,}000 \text{ utils}$$

$$x_0 = +\$900 \text{ million}$$

$$U_0 = +100 \text{ utils}$$

The parameters a and b which define the preceding utility function are calculated by solving the following equations:

$$U_0 = a + b\,\mathrm{Ln}(x_0 - x_{min})$$

$$0 = a + b\,\mathrm{Ln}(-x_{min})$$

Hence:

$$b = U_0 \,/\, \mathrm{Ln}(1 - x_0/x_{min}) = +72.134$$

$$a = -b\,\mathrm{Ln}(-x_{min}) = -411.44$$

Such an equation may or may not be applicable to real managers or companies making real decisions. It was chosen for its simplicity and capability to take into account the various aspects of a utility function that were discussed earlier.

The inverse of the utility function is the equation to be used to calculate the dollar value of a project when its utility $U(x)$ is known:

$$x = x_{min} + \exp[(U(x) - a)/b]$$

The equation $U(x) = a + b \, Ln(x - x_{min})$ used here is only one possible equation that can be used to represent a utility function. In this equation, a key assumption is that, however small the probability that a project will result in an NPV loss approaching the value x_{min}, this project will be rejected. This assumption is not necessarily valid. As mentioned earlier, there are methods that can be used to make such risky projects acceptable by changing the scope (e.g., phased investment) or sharing the risk (bringing partners). The possibility of a loss x_{min} might indeed be unacceptable, but mitigating actions can be used to reduce the magnitude of the possible loss to an acceptable level.

Some methods that are commonly used to determine the probability distribution of risk variables may result in rejecting projects for unsound reasons. If a normal distribution is assumed for the future price of the product to be sold, there is a nonzero probability that extremely low prices will be observed at some stage during the life of the project. These probabilities may be unrealistic. The results of any simulation study should be reviewed and analyzed for reasonableness, before conclusions are drawn that are driven by unrealistic assumptions. Methods for estimating utility functions without assuming a predetermined mathematical equation are discussed in the following section.

SIMPLE APPLICATIONS OF THE UTILITY FUNCTION

It was shown earlier that using the expected NPV to estimate the value of a project does not take risk aversion into account. It assumes the decision maker to be risk neutral. On the other hand, if the NPVs are converted to utilities, the expected utility takes into account the decision maker's risk aversion (or risk loving) as defined by the utility function.

A few simple examples are considered in this section that show how risk aversion changes the value that must be given to a project. The utility function U(x), where x is in millions of dollars, is assumed to be as follows:

If $x \leq -\$300$ million $\quad U(x) = -1{,}000$

f $x > -\$300$ million $\quad U(x) = -411.44 + 72.134 \, Ln(x - 300)$

TABLE 9.1 Application of the utility function—Example 1

	Probability of Occurrence p(x)	Net Present Value x (millions)	Utility U(x) (utils)
Outcome 1	50%	$100	20.75
Outcome 2	50%	$500	70.75
Expected value	100%	$300	45.75
Dollar equivalent	—	$300	$266

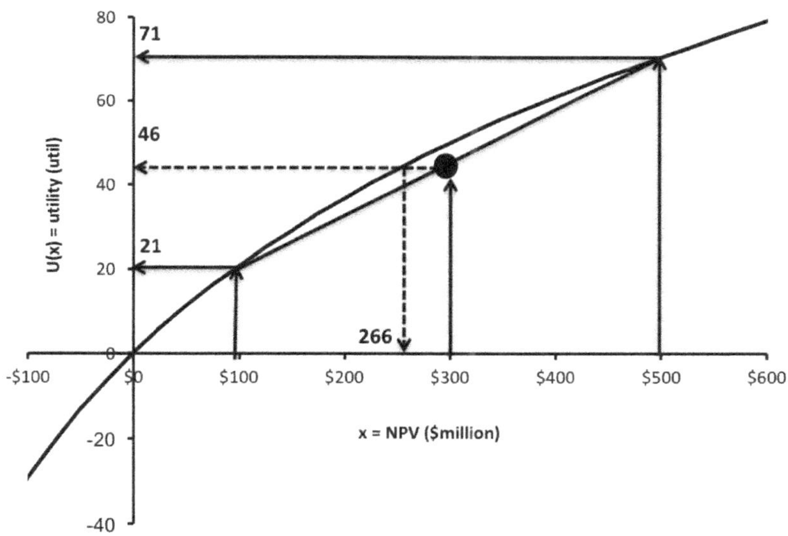

FIGURE 9.4 Estimating project value using utility function—Example 1

Example 1

Consider the example discussed at the beginning of this chapter, where there was an equal probability of a project's NPV being either $100 million or $500 million. The utility of this project and its guaranteed dollar equivalent is calculated as shown in Table 9.1 and illustrated in Figure 9.4.

The utility of $100 million is 20.75 utils; that of $500 million is 70.75 utils. The average NPV is $300 million, whereas the average utility is 45.75 utils.

TABLE 9.2 Application of the utility function—Example 2

	Probability of Occurrence p(x)	Net Present Value x (millions)	Utility U(x) (utils)
Event 1	50%	$200	36.85
Event 2	50%	–$200	–79.25
Expected value	100%	$0	–21.20
Dollar equivalent		$0	–$76

The dollar equivalent of 45.75 utils is $266 million, calculated using the inverse of the utility function:

$$x = -300 + \exp[(U(x) + 411.44)/72.134]$$

To the decision maker, this project with an uncertain return is worth the same as a guaranteed $266-million cash payment. This guaranteed dollar equivalent is not much lower than the $300 million obtained from averaging the NPVs. The reason for this small 11% difference is that the project value is positive whether Outcome 1 (NPV $100 million) or Outcome 2 (NPV $500 million) occur, and these positive values are both in the relatively flat part of the utility function, indicating no strong risk aversion in this range of values.

Example 2

In this and the following example, situations are considered where the risk of significant loss is increasing. The project shown in Table 9.2 is equally likely to result in a gain of $200 million or a loss of $200 million. The average NPV is zero, but the average utility is –21.20 utils, equivalent to a loss of $76 million. This implies that the company would accept a loss of $76 million to abandon the project and the uncertainty that comes with it. To avoid the risk of losing $200 million, the decision maker might decide to sell the project at a loss. If the project has a book value of $200 million, an argument could be made that the project should be sold for $200 – $76 = $124 million and that the remaining $76 million should be written off.

TABLE 9.3 Application of the utility function—Example 3

	Probability of Occurrence p(x)	Net Present Value x (millions)	Utility U(x) (utils)
Event 1	25%	$200	36.85
Event 2	75%	–$200	–79.25
Expected value	100%	–$100	–50.22
Dollar equivalent		–$100	–$150

Example 3

A more extreme situation is shown in Table 9.3. This project has 25% probability to result in a $200-million gain and 75% probability to result in a $200-million loss. Its expected NPV is a negative $100 million, while its expected utility indicates a risk-adjusted negative value of $150 million. Assuming a book value of $200 million, the decision maker could validly argue that the project should be sold for $50 million and the remaining $150 million should be written off.

It is relevant to note that while the project book value was $200 million to the seller, it will be $50 million to the buyer, which might be low enough for the buyer to decide to acquire the property. Differences between utility functions are also significant factors in changing the utility of a property from negative for the seller to positive for the buyer.

UTILITY FUNCTION AND DECISION TREE

How to represent and optimize multistage decision-making processes using decision trees was discussed in Chapter 7. An example was given that illustrated the use of decision trees to optimize early exploration projects. The corresponding decision tree, shown in Figure 7.2, is reproduced in Figure 9.5. The optimal decision was defined as that which maximized the expected NPV at each decision point. The decision was that we should go ahead with the exploration project since it had a positive expected NPV of $5 million. However, this decision did not seem reasonable since, even if both exploration and feasibility study were successful, there remained a 33.3% likelihood that the project would result in a $580 million loss.

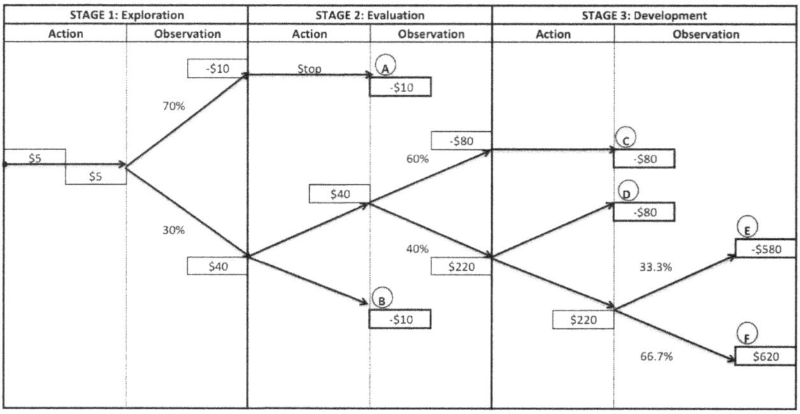

FIGURE 9.5 Decision-making using maximization of expected net present value

TABLE 9.4 NPVs and utilities used in original decision trees (Figures 9.5 and 9.6)

Input to Decision Tree	NPV (millions)	Utility (util)
U(A)	–$10	–2.45
U(B)	–$10	–2.45
U(C)	–$80	–22.37
U(D)	–$80	–22.37
U(E)	–$580	–1,000
U(F)	+$620	+80.83

Instead of using NPV as a measure of project value, one should use utility. In Figure 9.5, the dollar value of following each possible decision path is plotted in boxes A to F. The first step consists of replacing these dollar values by their equivalent utilities using the previously defined equations:

If $x > -\$300$ million $\qquad U(x) = -411.44 + 72.134 \,\mathrm{Ln}(x + 300)$

If $x \leq -\$300$ million $\qquad U(x) = -1{,}000$

The results are summarized in Table 9.4. Using utilities instead of NPVs, the optimum decision is made using the decision tree in Figure 9.6 following the process described in Chapter 7. If the exploration project was accepted as designed, its utility would be negative, equal to –2.45 utils. It should therefore be rejected or revised.

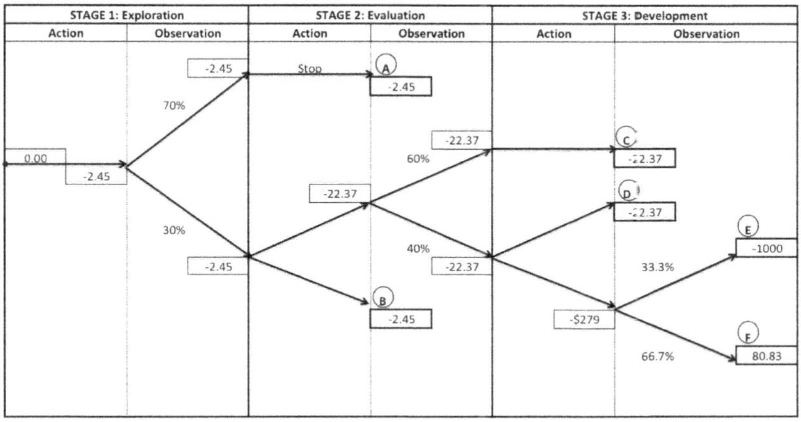

FIGURE 9.6 Decision-making using maximization of expected utility—original project design

The reason why the project should be rejected is that, even if exploration results are positive and the feasibility study ends with a recommendation to develop the mine, there is a very high likelihood (one chance in three) that the mine will have a negative cash flow resulting in a loss of $580 million. Before completely rejecting the project, one should look at modifying it, to significantly decrease the probability that the mine will fail if it is developed and to reduce the magnitude of the loss if it occurs. After considering several options, the following decisions were made by the project geologist:

- Stage 1 was redesigned to decrease the probability p(y1) that a feasibility study will be completed on an uneconomic project. The exploration cost C_1 was increased from $10 million to $20 million, resulting in improved geologic information on which to decide whether or not to go to Stage 2.
- As previously designed, the Stage 1 exploration project was expected to give positive results 30% of the time. This high percentage was the result of insufficient geologic information at the end of Stage 1 to justify abandoning the project. The risk associated with the project was not sufficiently reduced in Stage 1 and was transferred to Stage 2.
- As redesigned, the probability of positive results at the end of Stage 1 was decreased from 30% to 9% with the project being

considered for a Stage 2 feasibility study only if it had a high probability of success.

- The Stage 2 feasibility study was refocused to design a smaller operation that was less complex and requiring lower-cost infrastructure. Using improved information from the exploration Stage 1, the likelihood that the feasibility study would result in a recommendation that the mine be developed would be much higher, increasing from 40% to 90%.
- The Stage 2 feasibility study would be based on better information and would consider a smaller-size investment. The cost C_2 of the feasibility study would be significantly less, \$40 million instead of \$70 million.
- As a result of the smaller scale of operation, the cost C_3 of project development is expected to be \$175 million instead of \$300 million, and the cost of failure U_0 will be a loss of \$50 million instead of \$200 million. However, the expected NPV of the cash flow generated by the mine operation will be reduced from \$1 billion to \$750 million.

The new data is summarized in Table 9.5. The information to be entered in the decision tree is summarized in Table 9.6, and the decision tree based on utility rather than NPV is shown in Figure 9.7.

The probability that the exploration program will result in development of an uneconomic mine, with total loss estimated at \$285 million, is very low (1.2%). Consequently, the utility of the exploration program is positive, equal to 0.88 utils, and the new program should be approved. If NPVs had been used in the decision tree instead of utilities, the calculated value of the exploration program would have been \$22 million.

Because the utility of the exploration program is 0.88 utils, the dollar equivalent x of this utility can be calculated using the inverse of the utility function U(x):

$$x = -300 + \exp[(U(x) + 411.44)/72.134]$$

If U(x) = 0.88 utils, then x = \$4 million. In other words, the risk-adjusted value of the exploration program is \$4 million, whereas if risk is not taken

TABLE 9.5 Probabilities, costs, and NPVs of revised exploration program

Notation	Description	Value
	Probabilities (estimated by geologists and project engineers)	
$p'(U_1)$	Probability that there exists a deposit of value U_1	10%
$p(y_1)$	Probability that Stage 1 will give positive results	9%
$p(z_1\|y_1)$	Probability that Stage 2 will give positive results if Stage 1 is positive	90%
$p(y_1,z_1,U_1)$	Probability that a deposit of value U_1 will be discovered	8%
	Cost of three stages ($ millions, discounted to present)	
C_1 =	Cost of Stage 1, exploration	–$20
C_2 =	Cost of Stage 2, feasibility	–$40
C_3 =	Cost of Stage 3, mine development	–$175
	Net present value of project cash flow after development ($ millions)	
U_0	NPV of project if developed and it fails (does not include C_3)	–$50
U_1	NPV of project if developed and it succeeds (does not include C_3)	+$750

TABLE 9.6 NPVs and utilities of revised exploration program

Input to Decision Tree	NPV (millions)	Utility (util)
U(A)	–$20	–4.98
U(B)	–$20	–4.98
U(C)	–$60	–16.10
U(D)	–$60	–16.10
U(E)	–$285	–216
U(F)	+$515	+72.09

into account, the expected value is $22 million. If the owner of the exploration project considered selling it, offers between $4 and $22 million, depending on the buyer's risk aversion, could be expected; $22 million assumes no risk aversion, and $4 million assumes a level of risk aversion identical to that of the seller. The underlying assumption is that the buyer considers that he or she can complete the new exploration program as designed by the seller, at the same cost and with the same probability of success. If the buyer believes that success can be reached at lower cost and with higher probability, he or she will reach different conclusions concerning the value of the project.

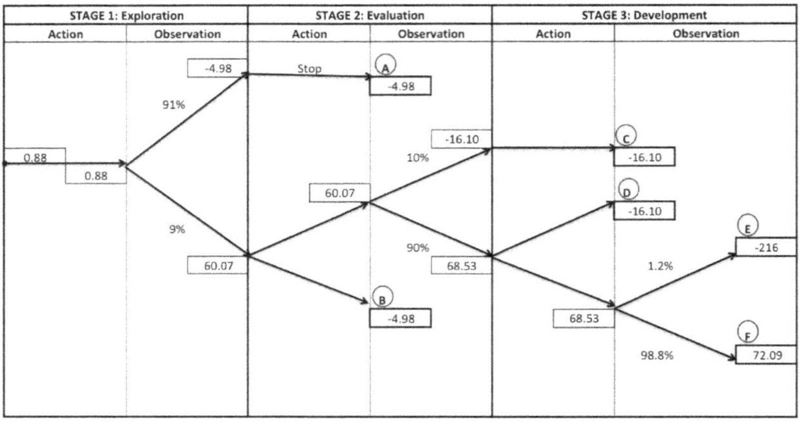

FIGURE 9.7 Decision-making using utility maximization with revised exploration program

HOW TO DEFINE A COMPANY'S OR INDIVIDUAL'S UTILITY FUNCTION

It was mentioned earlier that the model of utility function used here was chosen for its mathematical simplicity and its ability to reflect the expected properties of such a function. The function that represents the risk aversion of a specific company or decision maker at a given time and for a specific project could be significantly different. There is no simple way to define the utility function, especially given that risk takes different forms and the value of a project is not purely financial. A starting point would consist of assuming that most factors that define value can be summarized using monetary units. The following steps can then be followed:

1. Select an upper dollar value for the project under consideration, say, x_1 = \$900 million. The utility function will be calculated for values less than \$900 million.

2. Specify the utility of a project with assured value \$900 million, say, $U(x_1) = 100$ utils. This utility can be chosen arbitrarily. It does not influence in any way the decisions that will be made on the basis of utility.

3. Consider increasingly risky situations, where the project dollar value is expected to be x_1 = \$900 million with probability $p(x_1)$ or x_2 = \$0.0 with probability $p(x_2) = 1 - p(x_1)$. The utility of x_2 is $U(x_2) = 0$. The expected NPV of the project is

$$NPV = p(x_1)\, x_1 + p(x_2)\, x_2 = \$900\, p(x_1)$$

   This NPV represents the value of the project for a risk-neutral decision maker.

4. Consider a situation where $p(x_1) = 80\%$. The expected NPV is \$720 million. To determine the corresponding risk-adjusted value of the project, one should consider the following scenario and ask the following question:

   Consider a situation where the probability that the project will return \$900 million is 80% and the probability that it will break even (NPV = \$0) is 20%. What is the cash payment V you would accept in exchange for the project?

5. If the decision maker is risk averse, the answer should be less than the NPV of \$720 million, say, V = \$650 million. By definition, the utility of the project is U(V) which satisfies the following equation:

$$U(V) = p(x_1)\, U(x_1) + p(x_2)\, U(x_2) = 100\, p(x_1)$$

   The utility of a guaranteed \$650 million is 100 × 80% = 80 utils.

6. The same question should be asked for decreasing values of $p(x_1)$, from 100% to 0.0%. Table 9.7 gives an example of possible answers and how the utility function is defined accordingly. The results for positive NPV values are plotted on the right-hand side of Figure 9.8 (a continuous line). The horizontal axis represents the value of V in millions of dollars, and the vertical axis represents the corresponding utility.

The preceding steps are used to define the utility function when the cash equivalent V of a project is positive. The same process can be followed for negative cash equivalents. But first one must ensure that the scale used for negative NPVs is compatible with that for positive NPVs. To do this, one can use the following method:

TABLE 9.7 Estimation of cash equivalent and utility for increasingly risky projects with positive NPV

Project Description		Calculated NPV=	Cash Equivalent	Calculated Utility
x_1	$p(x_1)$	$x_1\ p(x_1)$	V	$u(V) = 100\ p(x_1)$
$900	100%	$900	$900	100
$900	80%	$720	$650	80
$900	60%	$540	$400	60
$900	40%	$360	$250	40
$900	20%	$180	$90	20
$900	0%	$0	$0	0.0

1. Select in Table 9.7 an NPV value for which the utility has been estimated. For example, x_1 = $650 million and $U(x_1) = 80$ utils.
2. Ask the decision maker what would be the maximum NPV loss x_2 that would be acceptable if there was a 50% chance that the project would have an NPV of either $650 million or x_2. The expected NPV would be $50\% \times 650 + 50\%\ x_1$. Increasing x_2 would result in giving preference to the project, whereas decreasing x_2 would result in abandoning it. The cash equivalent, and the utility of this project, would be zero.
3. For the purpose of this example, we will assume that the decision maker's answer is $x_2 = -\$200$ million. The utility of a project that offers an equal probability of a $650-million NPV gain or a $200-million NPV loss is zero.
4. The project utility is calculated as follows:

   $$50\%\ U(650) + 50\%\ U(-200) = 0$$

   The utility of a $200-million NPV loss is therefore

   $$U(-200) = -U(650) = -80 \text{ utils}$$

Once the utility of a $200-million loss has been estimated that is compatible with the utility given to positive NPVs, one can calculate the utility of negative NPVs between 0 and – $200 using the following approach:

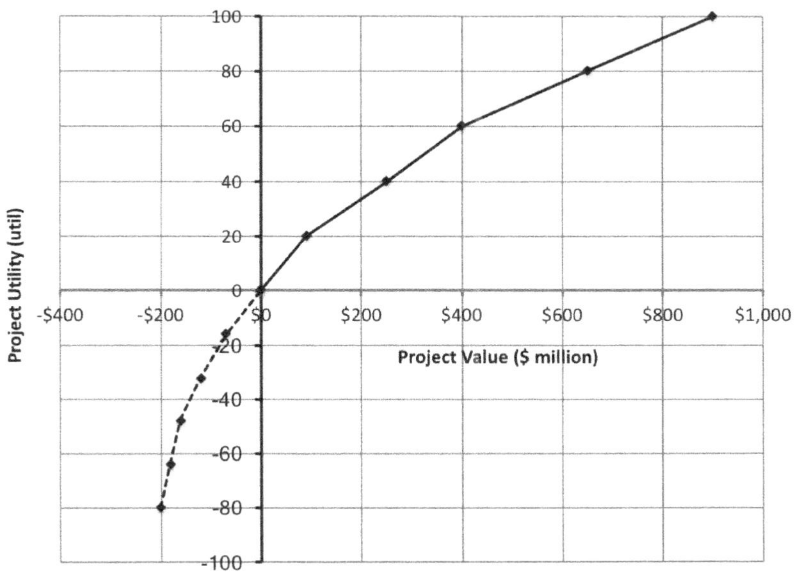

FIGURE 9.8 Utility function as a function of project cash equivalent

1. Consider a project whose NPV can be either $x_1 = \$0$ or $x_2 = -\$200$ million.

2. Consider increasingly more risky situations, where the project NPV value is expected to be $x_2 = -\$200$ million with probability $p(x_2)$ or $x_1 = \$0.0$ with probability $p(x_1) = 1 - p(x_2)$. The expected NPV of the project is

$$NPV = -\$200\ p(x_2)$$

3. To determine the corresponding risk-adjusted value of the project, one should consider the following scenario and ask a question such as this one:

   Consider a situation where the probability that the project will result in a loss of $200 million is 80% and the probability that it will break even is 20%. What is the maximum amount V you would accept to write off now rather than take the risk of losing $200 million with 80% probability?

TABLE 9.8 Estimation of cash equivalent and utility for increasingly less risky project with negative NPV

Project Description		Calculated NPV=	Cash Equivalent	Calculated Utility
x_1	$p(x_1)$	$x_1\ p(x_1)$	V	$u(V) = -100\ p(x_1)$
–200	100%	–$200	–$200	–80.00
–200	80%	–$160	–$180	–64.0
–200	60%	–$120	–$160	–48.0
–200	40%	–$80	–$120	–32.0
–200	20%	–$40	–$70	–16.0
–200	0%	$0	$0	0.0

4. The expected NPV is –$200 × 80% = –$160 million
5. If the decision maker is risk averse, the answer V should be between the maximum loss of $200 million and the expected loss of $160 million, say, V= –$180 million. The utility of a guaranteed $200 million write off is

$$U(V) = U(x_2)\ p(x_2) = U(-200)\ p(x_2)$$

$$U(-180) = -80 \times 80\% = -64 \text{ utils}$$

6. The same question should be asked for decreasing values of $p(x_2)$, from 100% to 0.0%. Table 9.8 gives an example of possible answers and how the utility function is defined accordingly. The results for negative NPV values are plotted on the left-hand side of Figure 9.8 (dashed line).

When using this empirical method to define the utility function, no assumption needs to be made concerning the maximum loss that the company might consider acceptable or the shape of mathematical equations that must be approximated. This approach is nevertheless overly simplistic in that it assumes that only one variable, the NPV, defines the utility function. But the principle of utility function as a representation of the risk-adjusted value of a project remains an integral part of a sound decision-making process.

CHAPTER 10

PROJECT UTILITY AND THE TRIPLE BOTTOM LINE

> *I may begin with the question of utility, which is a two-fold one, for either it may be asked whether the art of mining is really profitable or not to those who are engaged in it, or whether it is useful or not to the rest of mankind.*
>
> Georgius Agricola, *De Re Metallica*, 1556

The concept of "utility" is useful to show the need to incorporate more than financial performance in the evaluation of a project. While financial objectives play a critical role, other aspects, including environmental and social, must be taken into account. A historical review of the industry shows that many projects that were considered financially sound were delayed or canceled because of environmental or social factors whose significance was not properly assessed.

In traditional business accounting, the *bottom line* refers to profits or losses expressed in dollar terms. The social and environmental aspects are taken into account primarily because regulatory and legal obligations have a direct, quantifiable impact on the project's expected financial performance. An effective way to better define the utility of a project and to assess and mitigate the corresponding risks consists of using "full-cost accounting" and the "triple bottom line."[1] The three bottom lines measure financial (profit and loss), social (people), and environmental (ecological; the planet) performance, and are known as the three pillars of sustainability.

DEFINING THE PROJECT UTILITY USING FULL-COST ACCOUNTING

The economics of a project, its social and environmental impacts, and the confidence with which all three bottom lines are estimated are all factors that must be taken into account to estimate the project's utility and associated risk. The three bottom lines are not independent of each other. Ignoring the project's environmental and social impacts are likely to result in underestimation of the financial bottom line and in adding considerable risk to the project. The environmental impact would be minimized if we canceled the project, but this is usually not a viable option unless this impact is unacceptable and cannot be mitigated. The project's social benefit to the local communities would be maximized if all profits were assigned to social welfare, but this would imply ignoring legitimate expectations from shareholders and other stakeholders.

An often-used graphical representation of the triple-bottom-line framework is given in Figure 10.1. A circle represents each of the bottom lines. The circles overlap to illustrate dependence between bottom lines. The area where the three circles are intersecting is defined as the sustainable area. A company's objective should be to operate in the sustainable part of the graph, where the three bottom lines are appropriately taken into account and the financial, social, and environmental risks are assessed, balanced, and controlled.

The United Nations defines *sustainable development* as "development that meets the needs of the present without compromising the ability of future generations to meet their own needs" (UN 1987). While this definition is applicable to the mining industry, its meaning needs to be clarified. What is meant by "the needs of the present and future generations"? The needs and expectations of the shareholders, employees, suppliers, and clients are

1 "The phrase 'the triple bottom line' was first coined in 1994 by John Elkington, the founder of a British consultancy called SustainAbility" (*The Economist* 2009).

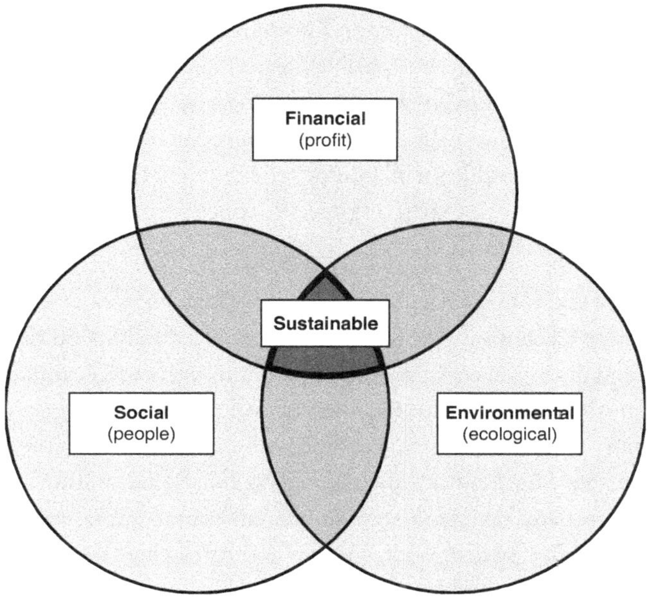

FIGURE 10.1 The three pillars of project sustainability

different and often conflicting with those of the country or region where the mine is located. High-grading a deposit may be necessary to reduce risk and satisfy the current need for economic viability and financial sustainability, but it may be compromising opportunities for future generations.

There are many stakeholders who are interested in the performance of a project and who must be taken into account when assessing the three bottom lines. A list of stakeholders can be as follows:

- Shareholders
- Banks and financial institutions
- Business analysts
- Employees and their families
- Customers
- Suppliers
- Local communities
- Local, regional, federal, or country governments
- Future generations
- Nongovernmental organizations (NGOs)

The value given to the three bottom lines varies with each stakeholder. For a project to operate in the sustainable area, a solution must be found that involves trade-offs between many competing and sometimes conflicting interests. Solutions must be found that recognize the concerns of each stakeholder. It is rarely possible for all interests to be completely satisfied. The risk that may be taken by not satisfying specific concerns must be fully assessed and accepted or mitigated.

The *environmental* (also called ecological or planet) *bottom line* takes into consideration the impact that the project is expected to have on the planet's physical and biological environment. Most of this impact is local, at the mine site and in surrounding areas. Regional environmental impact can also take place along access routes. The stakeholders who are interested in the environmental bottom line are local, but also regional, national, and international. Failure to properly take into account the environmental bottom line can result in loss of license to operate, not only at the site where this failure occurs but also at other sites worldwide.

The *social bottom line* is a function of the impact, positive or negative, that the project is likely to have on people. This impact is worldwide, as the performance of a major project is of direct interest to company shareholders, NGOs, central (country) government, regional and local governments, local communities and the organizations representing them, and individuals, especially those directly and personally affected by the project. As is the case with the environmental bottom line, failure to recognize the social impact of the project on one or more of the stakeholders can have dire consequences on both the project and the ability of the company to operate worldwide.

The utility of a project increases when the three bottom lines converge toward sustainability. High confidence that the project falls within the sustainable area should be a requirement before deciding that a reserve can be declared and a project can be developed.

There are many input variables that inform the decisions made in each stage of project development. Full-cost accounting defines a logical framework to analyze the nature of these variables, grouping them according to whether they influence the financial, social, or environmental bottom line. Many input variables influence more than one bottom line. These cross-correlations significantly increase the complexity of project evaluation, risk assessment, and risk mitigation. To manage project risk, one must understand the influence of each variable on all three bottom lines simultaneously. Failing to do so will result in underestimating project risk. The influence of one input variable on one bottom line is likely to be less, even significantly less, than the influence of the same input variable on the project as a whole. Variables that influence the bottom lines are analyzed in Chapter 11.

MEASURING THE THREE BOTTOM LINES

When evaluating projects, companies use processes that include all aspects of full-cost accounting but might not formally recognize the three bottom lines as separate but related entities whose values must be estimated. When investing in a project, the risk that is taken depends not only on the expected value of the three bottom lines but also on the uncertainty with which these values are estimated. All stakeholders must be considered. Each stakeholder is likely to have a different opinion of the magnitude of the impact that the project will have on each bottom line, and whether this impact is acceptable. Even if there is agreement on financial measures, such agreement is not likely when considering the social and environmental bottom lines.

To measure anything, one must agree on a unit of measurement and the process to be followed to quantify the results. There is no common measure applicable to all three bottom lines. Profits are measured in dollars, but social and environmental impacts are not necessarily quantifiable in dollar terms. How do we measure the change in quality of life that results from displacement of a population to obtain access to land needed for the project? How do we value loss of aquatic biodiversity from swamps located over a future mine site? Full-cost accounting requires assessment of the three bottom lines, but how can we combine these bottom lines to assess the value of the project as a whole and the risk associated with it? How do we decide that the project falls within the sustainable area shown in Figure 10.1?

One may attempt to monetize all the dimensions of the triple bottom line. While this would have the benefit of using a common unit, it is unlikely that agreement can be reached on the dollar value of social well-being or environmental damage. These values are highly subjective, and each stakeholder, looking at it from a differing point of view and using a different value system, is likely to have a different opinion. A method occasionally used to mitigate social or environmental impact in one location consists of making available compensating opportunities in another location. Although the cost of such mitigation may be quantifiable, it is not always available and may result in replacing one impact with another of a different type in a different location.

Monetizing the three bottom lines is not always feasible. A solution might consist of using indexes to reflect the value of the bottom lines, thus eliminating the incompatible unit issue. If there were a universally accepted indexing method, comparison between projects or between companies would be possible. But there is not yet agreement on such a method, and subjective judgment is unavoidable even—and possibly more so—when using indexes. Different stakeholders will give a different index value to the social welfare or environmental impact of the same project. How to compare the index value of each of the three bottom lines and decide which is acceptable remains an elusive question.

Because of these complexities, an effective way to measure the three bottom lines and the risk associated with them consists of using separate measures but one single graphical representation of the results obtained. The mining company needs to determine which variables should be considered to measure each of the bottom lines, the weight to be given to each variable, and how to combine their impact to obtain a simple, widely understood, and accepted index value.

GRAPHICAL REPRESENTATION OF BOTTOM-LINE UNCERTAINTY

A process that consists of considering the three bottom lines separately has significant weaknesses. While it is true that mining decisions influence each bottom line in various ways, improving estimation of one of the bottom lines is likely to influence, positively or negatively, the other bottom lines. For example, reducing the risk associated with the social impact of the project may require improving water supply to surrounding communities, which results in additional costs, which influences the financial bottom line. Building dams to create a stable water supply to a milling operation will have

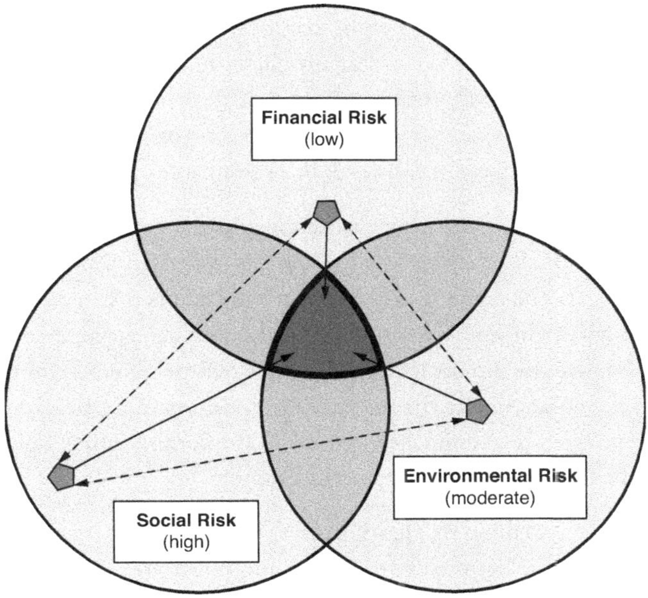

FIGURE 10.2 Graphical representation of risk associated with, and relationship between, each bottom line

environmental and social impacts and will require an environmental study to get the necessary permits.

The uncertainty associated with current knowledge of the value of each bottom line can be represented as shown in Figure 10.2. At any given time, the project has a given likelihood of falling within the sustainable area. The level of risk associated with each bottom line is a function of the probability that the bottom line will fall within the sustainable area. In Figure 10.2, the position of the project bottom lines with respect to the sustainable area is represented by small gray polygons. The farther the polygons are from the center of the figure, the higher the probability that the bottom line will not fall within the sustainable area. The solid arrows represent the distance between the current estimate of each bottom line and the sustainable area. The dashed arrows illustrate the interrelationships between bottom lines.

To be useful, the distance between each polygon and the sustainable area must reflect the uncertainty with which the corresponding bottom line is known and the risk to the project that results from this uncertainty. Such a distance is indicative of the attention that must be given to each bottom line to ensure that the project will eventually fall within the sustainable domain. The appropriate risk level is obtained when all three polygons fall within the target area. A project deserves further consideration only if there is some likelihood that its development will result in a sustainable operation.

There are a number of significant advantages in using such a graphical representation of the level of knowledge one has at a given time concerning a mining project. The graph is easy to understand across disciplines. It clearly shows the areas where risk is most significant and management attention is needed to move the project forward. The units of measurement of the three bottom lines do not need to be the same. The position of the polygons is unit free: It represents the confidence one has in the corresponding bottom line and whether this level of confidence is acceptable.

It should be kept in mind that getting better information does not necessarily reduce the distance between bottom line and target area. There can be circumstances where additional information shows that the risk is higher than previously estimated or that it cannot be managed to the satisfaction of management or the stakeholders. In such a situation, the project might have to be abandoned.

Figure 10.2 illustrates a situation likely to be found in the early stage of project evaluation. Detailed geological and technical information has been obtained, some limited environmental assessment was completed, and the potential social impact and how to manage it has not yet been analyzed. The figure shows that the financial risk is limited except that considerable uncertainties remain concerning environmental and social impacts, and the potential cost of mitigating these impacts. The environmental risk is significant, but the uncertainty associated with the social aspect of the project is extremely high. This graphical representation helps prioritize resources as needed for risk management.

ORGANIZATION CHART FOR FULL-COST ACCOUNTING

The nature of the three bottom lines makes it difficult to quantify and compare them on the same value scale. This justifies analyzing each bottom line separately. Different expertise is needed, and communication between experts is challenged by the use of different expert languages. The problem is exacerbated by company organizations, which group experts in different departments, and with different reporting lines. The geology department, who supplies the most critical information needed to evaluate a project, may report directly to the chief executive officer. Technical functions, including mining, processing, project engineering, and project management, are likely to be part of a technical services department reporting to the chief operating officer; the head of environmental affairs may report directly to the chief executive officer; community relations may report to the head of external relations. A separate risk and compliance department may be responsible to evaluate the overall project risk.

Figure 10.3 shows a possible organization chart for full-cost accounting. Individuals from the different company departments are assigned to the project on a full-time or part-time basis as the need arises. It is the responsibility of the project manager to ensure that he or she has access to the right personnel and financial resources at the right time. This cannot be done effectively without understanding and support from the company's senior management. Failure to do so can significantly increase project risk.

One might assume that communication between departments is more effective in smaller companies. But this is not necessarily the case, as communication is a function of language and personality, as much it is of organization charts and systems. The likelihood of failure in information sharing and transfer between departments is likely to increase with the complexity of the company's organizational structures.

Matrix management has been promoted as an effective way to improve communication. However, this is only the case if the lines of responsibility are clearly defined. The manager of a technical services group may have a number of metallurgists reporting directly to him or her. At any given time, a metallurgist may be seconded to a project and have a dotted line of responsibility to the project manager. During this secondment, the primary responsibility of the metallurgist should be toward the project manager. The main role of the technical services manager should be to assign resources, to ensure quality of services, and to manage conflicts when necessary. This role becomes increasingly complex when resources become scarce and the same metallurgist is assigned to more than one project. If projects are to be successfully completed, both management

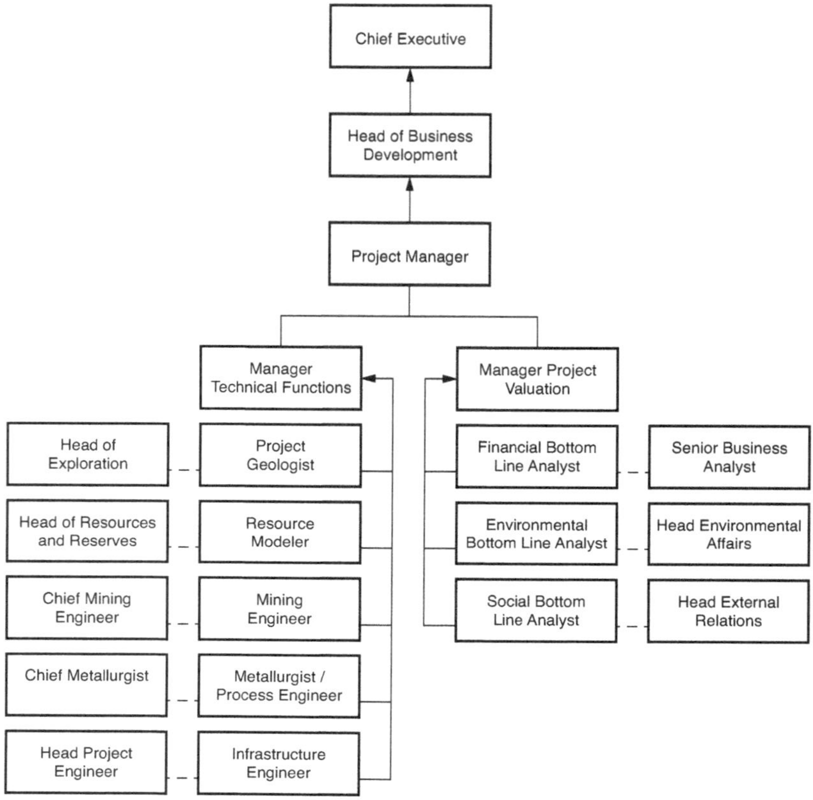

FIGURE 10.3 Typical organization chart for evaluation of mineral projects

and technical staff must have a clear understanding of the boundaries within which they operate.

The ability to properly assess total project risk falls under the responsibility of a project manager, who must have appropriate understanding of each of the project components and how they relate to each other. The project manager must ensure appropriate communication among the different departments, taking into account that such communication requires a common language that is often lacking, resulting in misunderstandings, incorrect decisions, or lack of necessary actions.

HOW TO EVALUATE THE THREE BOTTOM LINES

The utility of a project, and whether it is understood with sufficient confidence to publish a reserve estimate and commit to development of a new mine, is a function of the value of the three bottom lines (financial, environmental, and social) and the confidence one has in these values. Although significant overlaps exist, evaluation of each bottom line requires different areas of expertise. The tendency is to evaluate each bottom line separately. However, effective project management requires combining all bottom lines into one project utility. The ability to do so requires specialized expertise, including significant experience in estimating and understanding all factors that constitute value and risk.

A process that a project manager can follow to assess the value of a project and the uncertainty associated with this estimated value is as follows:

1. Consider the three bottom lines separately.
2. Assign a "bottom-line analyst" to each bottom line (Figure 10.3). The analyst's responsibility is primarily to ensure that the bottom line is estimated correctly, including uncertainty.
3. Qualifications required of a bottom-line analyst include proven expertise in understanding all the risk factors potentially affecting the bottom line, the ability to effectively communicate with the expert team for which he or she is responsible, knowledge of which information is needed for appropriate bottom-line estimation, and the ability to obtain this information wherever it may be located.
4. The bottom-line analyst must ensure that all variables that influence the bottom line have been correctly identified. These variables are defined as "input variables." Complete identification of these input variables is critical. Many projects fail or underperform because the evaluation team was not aware of some critical risk factor, critical input variables were not properly estimated, or an early decision was made to ignore some input variables without properly assessing their significance.
5. Once the input variables have been identified, their values must be determined, as well as the confidence with which they are known.

The value of a variable may not be a specific number but rather a level of understanding of the current situation and how this situation may change as the project progresses. For example, one may not be able to quantify the emotional as well as socioeconomic impact of relocating landowners, but one can assess how such relocation should be approached to minimize this impact, taking into account short- and long-term consequences as well as local, regional, and worldwide repercussions.

6. Correlations between input variables must be identified, not only as they influence a specific bottom line but also how they are likely to influence other bottom lines. A positive change in one input variable can have a positive or negative influence on the value of another input variable. The list of input variables that influence one bottom line should be compared with that influencing the other bottom lines.
7. The experts responsible for estimation of each input variable must determine the level of uncertainty associated with these estimates.
8. The relationship between the value of the input variables and the bottom line must be appropriately modeled. The sensitivity of the bottom line to each input variable must be determined, taking into account the level of confidence that one has in the input variables. Only then is it reasonable to decide that some of the input variables do not need to be given additional attention because their impact on the project value is not significant.
9. Input variables that have a significant impact on the bottom line should be ranked. The need or ability to improve understanding of these variables should be evaluated. Mitigating actions should be identified, and the costs and benefits of these actions should be quantified. These costs and benefits should take into account all bottom lines that are influenced by these variables.

During this entire process, the project manager must ensure that there is effective communication between bottom-line analysts to avoid underestimation of the project risk, especially when input variables influence more than one bottom line. At any given time, and especially at the end of each stage of project development ("stage gate"), the results obtained should be plotted in a format such as that shown in Figure 10.2. If one or more of the bottom lines are outside the sustainable area, determination should be made of what actions could be taken to improve this bottom line, and the cost of such actions. Recommendations should be made on how to proceed. This may include recommendation to get missing information, improve the analysis of available information, put in place risk-mitigating actions, delay the project, sell it, or abandon it.

If there is high confidence that the three bottom lines fall within the sustainable area, the recommendation should be to publicly declare a mineral reserve and to go ahead with project development.

RELATIONSHIP BETWEEN INFORMATION AND BOTTOM LINES

The relationship between information and bottom lines is complex. Many input variables influence more than one bottom line. These cross-correlations significantly increase the complexity of project evaluation and management, as well as risk assessment and mitigation. Risk management requires understanding the influence that each variable has on all three bottom lines simultaneously. Failing this, risk will be underestimated, possibly significantly. In Chapter 11, each bottom line and the variables that influence it are considered separately. This approach recognizes that different expertise is needed to evaluate each bottom line and that company organizations reflect this need. However, variables that are listed as influencing one bottom line may also be listed under another bottom line.

Another approach to analyzing risk assessment consists of classifying input variables according to the technical departments whose decisions contribute most to defining the project. This approach is the subject of later chapters in which the role of the departments responsible for exploration, resource estimation, mining, processing, and infrastructure is discussed separately.

The technical departments collect data, complete studies, and generate information that is needed to value the three bottom lines. Such information includes deposit models, mine plans, process designs, production schedules,

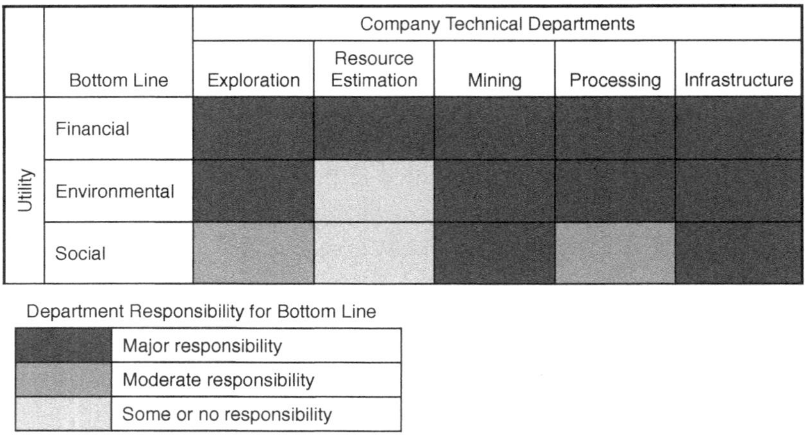

FIGURE 10.4 Relationship between technical departments and the three bottom lines

capital and operating costs, and expected cash flows. The likely relationship between technical departments and bottom lines is illustrated in Figure 10.4, the darker cells representing stronger relationships.

Other departments play a critical role in submitting information that is needed for the technical departments to complete their tasks and for the project manager to fully evaluate the three bottom lines. The scope of such information includes legal, safety, environmental, financial, taxes, information technology, human resources, sales and marketing, procurement, and logistics. The role of these departments is taken into account implicitly (but not explicitly) in this book, which focuses on technical aspects.

THE RISK REPORTING MATRIX

The significance of a risk variable, and whether or not it deserves further consideration, is a function of three parameters:

1. The values that this variable may take when the project is developed
2. The probability that such values will be observed
3. The impact that such values will have on the bottom lines

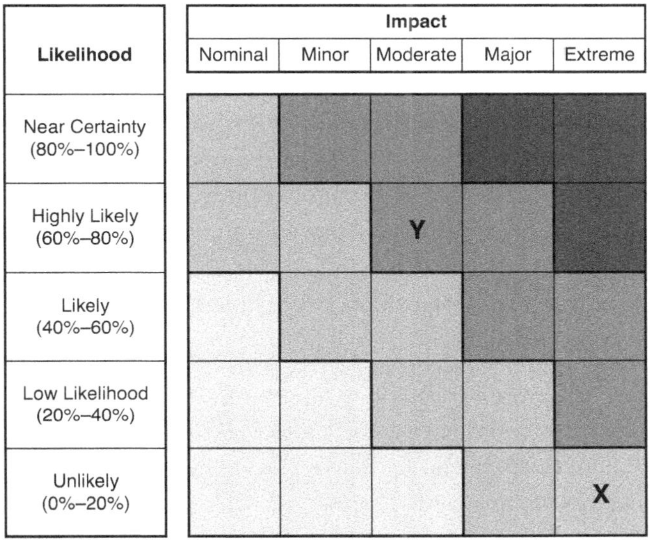

FIGURE 10.5 Risk reporting matrix showing classification of two risk variables, X and Y

The *risk reporting matrix* is an easily interpreted graphical method that can be used to illustrate the contribution to project risk that can be directly attributed to uncertainty concerning a specific input variable (Figure 10.5). In such a matrix, the vertical direction reflects the likelihood that an input variable will have a given value or that an event will occur during the life of the project. The horizontal direction reflects the impact that this variable will have on the project value. The matrix is divided from lower left to upper right in different sectors, which reflect increasing risk. A first step in designing a risk matrix consists of defining which variables could have a nominal, minor, moderate, major, or extreme impact on the project.

Variables that present a negligible likelihood of even a moderate impact may not deserve further attention. Conversely, a variable that could have an extreme impact may require immediate consideration even if the probability of occurrence is very low. For example, there might be a very low probability (less than 10%) of a major earthquake during the life of the project, but the impact of such an earthquake might be catastrophic. The letter X on Figure 10.5 represents the corresponding risk. Although this risk is considered moderate, mitigating actions should be taken in infrastructure design and emergency planning.

Another example is that of a deposit whose geology is believed to lack complexity but about which there is limited information. It might be highly likely (75% probability) that the properties of the deposit will differ from those currently expected, but the impact may only be moderate. The letter Y on Figure 10.5 represents the corresponding risk. This risk is considered high, and additional geologic information should be acquired before the project is approved for development.

The risk reporting matrix is useful to illustrate the impact of individual variables or groups of correlated variables, such as "geology," "metallurgical process," or "social environment." However, it is of limited use for variables that can take a range of values with different probabilities or where there are complex cross-correlations between variables. It is then necessary to use different methods, such as those described in Chapter 8.

CHAPTER 11

VARIABLES INFLUENCING THE THREE BOTTOM LINES

A miner, before he begins to mine the veins, must consider seven things, namely: the situation, the conditions, the water, the roads, the climate, the right of ownership, and the neighbours.

Georgius Agricola, *De Re Metallica*, 1556

Input variables are data that inform the decisions made in each stage of project evaluation. Full-cost accounting defines a logical framework to analyze the nature of these variables, grouping them according to whether they influence the financial, social, or environmental bottom line.

VARIABLES INFLUENCING THE FINANCIAL BOTTOM LINE

The financial bottom line is measured in dollar terms, and its value is assessed from the cash flow that the project is expected to generate during its entire life. A number of measures are used, including net present value (NPV), internal rate of return (IRR), yearly cash flow, capital requirements, payback period, and so forth. The weight given to each measure varies with the company's strategic objectives, its ability to finance new projects, its appetite for risk, and other priorities as defined by senior management.

General Considerations

The financial bottom line takes into account production schedule, capital and operating costs, and revenues from sales as well as taxes, royalties, and other variables that influence cash flows. Sources of uncertainty include insufficient or inadequate information concerning geology, in-the-ground mineral resources, and how much of these resources can be mined, processed, and sold at a profit. There is uncertainty concerning capital and operating costs. The future value of the salable product may not be known. Confidence that mine and processing plant will perform as planned may be limited. Environmental, social, political, legal, and regulatory factors can play a significant role in the financial bottom-line risk assessment.

The financial bottom line is estimated from information supplied by all members of the project team. The project manager is responsible for all three bottom lines and must ensure that the analyses completed by all members of the team are sound. Confidence in the financial bottom line depends on the following:

- Integrity and absence of conflicts of interest among those who evaluate the financial bottom line (For example, a manager may be promoted if the project moves forward; a consultant may give results that are biased by company expectations; or an engineering, procurement, and construction company may see an opportunity to obtain a very lucrative construction project.)
- Competence of the financial analyst responsible for evaluation of the financial bottom line
- Competence of individuals supplying information needed to estimate the financial bottom line
- Whether all variables that influence the bottom line have been properly identified and estimated
- Whether the uncertainty of the estimates has been quantified
- Whether the model used to estimate the financial bottom line, including the life-of-mine cash-flow model, takes into account all relevant input variables and combines them appropriately
- Whether the model used to quantify the risk to the bottom line takes into account both space- and time-related uncertainty

The life-of-mine cash flow that the project is expected to generate is estimated from the value of key input variables. Some of these variables influence the bottom line directly, while others must be considered because of their indirect influence. Capital costs, unit operating costs, taxes, and dues have a direct impact. Tons mined, and quantity and quality of material processed and sold have an indirect impact, as they must be combined with unit costs, recoveries, sales prices, and so on, to be converted into dollar terms. Understanding of both direct and indirect contributing factors is necessary to quantify the financial bottom line and the uncertainty with which it is known.

Project documentation, including documentation of assumptions made and their justification, should be a major part of the company's learning process. Analysis of past projects, including those that succeeded as well as those that failed, represents an important source of information and decreases the likelihood that past mistakes will be repeated.

Sunk Costs and the Financial Bottom Line

Sunk costs are expenses that were incurred in the past and will not be changed by future decisions. The decision whether a reserve can be declared and a mine can be developed is independent of sunk costs. Whether costs incurred in the past were appropriate or excessive should have no influence on decisions to be made in the present. Excessive past expenditures do not necessarily imply that past decisions were inappropriate. Decisions may have been made assuming values of key variables such as metal prices, political environment, or social acceptance, which may have appeared reasonable at the time but were proven wrong at a later date. Alternatively, management may have taken excessive risk or been overly optimistic or misinformed. Reasonable precautions may not have been taken to mitigate risks that were known at the time expenditure decisions were made.

Sunk costs that have been capitalized are taken into account when calculating taxes and determining whether assets need to be written off. A manager who was responsible for excessive expenditures may be reluctant to accept the need to write off these expenditures, as it may reflect poorly on his or her ability to make sound decisions. Subtle, and occasionally not so subtle, pressure may be placed on the project team to predict cash flows that are sufficient to avoid a write-off. Such cash flows may be overly optimistic and such actions would increase the risk that the project will fail. Such behavior would be unprofessional, unethical, and, in extreme cases, illegal.

Technical Information and the Financial Bottom Line

A short list follows of technical information that is needed to estimate capital and operating costs during the expected life of the project. These technical requirements are described in detail in subsequent chapters. Information is usually required on a yearly basis or for shorter time periods, quarterly or monthly, during early years.

- Mining infrastructure and investments to be built and capitalized (shaft sinking, haulage drifts, underground development, pre-stripping, etc.)
- Mining method and productivity
- Processing infrastructure and investments to be capitalized (processing plant, tailings storage facilities, laboratories, etc.)
- Processing method, plant recovery, and productivity
- Other infrastructure requirements and design, including offices, housing, dams, roads, rail facilities, harbors, and so on
- Tonnage and quality of material scheduled to be mined on a yearly basis, classified by destination to waste dump, stockpiles, processing plant, and leach pads
- Tonnage and quality of material recovered from stockpiles and delivered to the processing plant
- Tonnage and quality of material processed
- Quantity and quality of salable product recovered and available for sale
- Quantity of rejects to be sent to tailings facilities
- Expected time to acquire necessary permits

- Time needed to build the project infrastructure after necessary permits and corporate approval are obtained and until planned commercial production is reached
- Closure and environmental remediation technical requirements

Financial Information and the Financial Bottom Line

Financial information is that which is needed to estimate capital costs, operating costs, and annual cash flows, taking into account technical requirements. This information must be estimated on a yearly or shorter basis, for the life of the project:

- Expected costs to acquire permits yet to be obtained
- Initial capital cost (mining, processing, infrastructure, and capitalized mine development)
- Contracted and owner's costs to be included in capital and operating costs
- Sustaining capital
- Mine operating costs per ton of material, classified by origin and destination, including drilling, blasting, loading, hauling, ore control, and waste and stockpile management
- Mine maintenance costs
- Mine overhead costs
- Processing cost per ton of material, including plant operating costs, laboratory costs, and tails disposal
- Processing maintenance costs
- Processing overhead costs
- Market value per unit of product sold
- Cost of sales

- Reclamation costs (continuous reclamation, end-of-life closure, and costs in perpetuity)
- Yearly site and corporate overhead costs
- Exploration costs attributable to project (future costs only)
- Taxes, duties, and royalties
- Debt servicing
- Environmental and social costs as listed below
- Exchange rate, inflation, and deflation as expected over time

Environmental Information and Financial Bottom Line

The project manager must ascertain that appropriate attention was given to the environmental impact, taking into account legal and regulatory obligations, company policies and guidelines, ethical values, and expectations from stakeholders. Environmental variables that will influence the financial bottom line include

- Time needed to develop the legally required environmental impact statement (EIS), including appropriate mitigation plan;
- Time needed to obtain all necessary permits;
- Yearly environmental impact mitigation costs, to be included in the cash flow either as part of mining, processing or overhead costs, or as stand-alone line items; and
- Environmental closure costs at the end of the mine life and costs to perpetuity.

Social Information and the Financial Bottom Line

The project manager must ascertain that appropriate attention was given to the social impact. Social variables that will influence the financial bottom line include

- Time needed and costs of obtaining baseline social information and estimates of project social impact;
- Time needed and cost of developing required social impact documents, including appropriate mitigation plan, and to obtain necessary permits;
- Initial cost of social impact, including cost of land acquisition, relocation, housing, schools, medical facilities, training, and infrastructure improvement;
- Yearly social impact mitigation costs; and
- Social closure costs at the end of the mine life to facilitate transition to a postmining economy.

Uncertainty and Risk Assessment

The financial bottom line is estimated using input variables whose values are likewise estimates. The confidence with which the financial bottom line is estimated is a function of the confidence one has in the input variables. It is also a function of the confidence one has in the model used to estimate key financial performance indicators from the input variables. Is the model structure complete? Does it consider all inputs, effects, and relationships? Does it accurately represent the relationship between input and project value?

The risk factors most likely to impact the financial bottom line include the following:

- **Geology:** Likelihood that the deposit properties, including tonnage, grade, mineralogy, and geotechnical and hydrological characteristics, will be as indicated by the resource model
- **Schedule, budget, and scope:** Ability to build the project on schedule and within budget without change of scope
- **Ramp-up schedule:** Delays in bringing mine and processing plant to commercial operation, and in generating a positive cash flow, can significantly decrease the project's net present value

- **Mine productivity:** Ability to deliver the planned tonnages and grades of material to the processing plant, at the estimated cost
- **Mill productivity:** Ability to process the tonnages delivered and to recover the planned amount of salable product, at the estimated cost
- **Market risk:** Can the market absorb expected production? Will the value of product sold be as assumed in the cash-flow analysis?
- **Political risk:** Risk of changes in laws, regulations, and taxes; risk of creeping or outright nationalization
- **Environmental risk:** Change in stakeholder expectations; failure of environmental impact mitigation plan
- **Social risk:** Change in stakeholder expectations; failure of social impact mitigation plan
- **Catastrophic natural events** (tsunamis, earthquakes, fires, floods, droughts, snowstorms, etc.)

The risk associated with a given mine plan is quantified by analyzing the resulting production schedule when different assumptions are made concerning the simulated value of the input variables. Methods that can be used to estimate this risk were discussed in Chapter 8. These methods can be classified in the following order of complexity.

1. **Base case:** The value of each input variable is defined by its "best estimate." The financial bottom line is calculated using these estimates. Only one value is obtained, which may or may not be the most likely value. All project studies start with development of a base case. Because risk is not quantified, this information is only sufficient in the simplest situations.

2. **Sensitivity analysis:** The value of each input variable is changed linearly by some percentage (e.g., 10%) to determine their influence on the bottom line. This sensitivity analysis is useful to determine which variables can add significant risk to the project. Sensitivity analyses do not estimate likelihood of occurrence and cannot be used to quantify risk.

3. **Risk analysis without flexibility:** The probability distribution of the input variables is estimated, both in space and over time. Monte Carlo simulation is used to randomly specify the value that each variable may take. A single realization of the financial bottom line is calculated for each simulation. The probability distribution of the financial bottom line is determined by calculating possible realizations for a number of input data simulations. In these calculations, it is assumed that there is no flexibility to the mine plan. While overestimated, the calculated risk gives useful information and may be sufficient to make final decisions.
4. **Risk analysis with flexibility:** The same process is used as in the previous case, except that future early cash flows are estimated taking into account management's ability to change the plan when new information becomes available. Such changes are usually made at the end of each year of operation, taking into account the simulated values of the input variables up to the year under consideration (but not beyond this year). This approach gives the most realistic estimate of the project risk.

The results of this risk analysis depend on the base-case assumptions, including mine design, processing capacity, and derived capital and operating costs. The initial assumptions are not likely to be the best possible ones. In most cases, it is advisable to consider one or more different base cases or scenarios and compare the corresponding project values and risks before making a final decision.

VARIABLES INFLUENCING THE ENVIRONMENTAL BOTTOM LINE

> *The strongest argument of the detractors [of mining] is that the fields are devastated by mining operations, for which reason formerly Italians were warned by law that no one should dig the earth for metals and so injure their very fertile fields, their vineyards, and their olive groves. Also they argue that the woods and groves are cut down.... and when the woods and groves are felled, then are exterminated the beast and birds, many of which furnish a pleasant*

> *and agreeable food for man. Further when the ores are washed, the water which has been used poisons the brooks and streams, and either destroys the fish or drives them away.*
>
> Georgius Agricola, *De Re Metallica*, 1556

Legally required environmental assessments and environmental impact statements are expected to accurately assess both the physical impact of a project on nature and the environment, and the social impact on people and society. Indeed, while not fully overlapping, environmental and social impacts are strongly linked. However, different expertise is needed to assess the impact of a project on people and on the natural environment. In the triple-bottom-line approach to project evaluation, these two types of impacts are analyzed separately. Provided the study of each bottom line is consistently coordinated with that of the other bottom lines, this approach reduces the risk that some aspects of the overall value of the project are overlooked or neglected.

General Considerations

The environmental bottom line takes into account rules and regulations that must be followed to obtain necessary permits, as well as often-higher ethical and behavioral standards that the company expects its employees to follow. Reliable estimation of the environmental bottom line depends on the following:

- Competence of the analyst responsible for evaluation of the environmental bottom line; competence of persons supplying information to the environmental analyst as needed to evaluate the environmental bottom line
- Knowledge of applicable rules, regulations, and environmental permitting requirements; understanding of company objectives, values, and ethics concerning protection of the environment
- Understanding of the likely environmental expectations from local communities and nongovernmental organizations (NGOs) (These are likely to differ from and be more taxing than regulatory requirements and company expectations.)

- Whether all variables that influence the environmental bottom line have been properly identified and estimated
- Whether the uncertainty of the estimates has been quantified
- Whether the uncertainty concerning environmental impact and short- and long-term consequences has been quantified
- Whether environmental impact remediation plans have been developed and evaluated
- Whether the process used to estimate the environmental bottom line takes into account all relevant input variables and combines them in a way that fairly represents the company's objectives and values
- Whether the expected environmental impact has been adequately presented to and discussed with all stakeholders (Effective communications are required to get acceptance of this impact.)

A complete baseline environmental survey of the project site and surrounding areas, including expected location of required lines of communication (roads, railroads, harbor, power lines, pipelines, etc.) must be completed before initiating significant on-site activities. An appropriate environmental impact mitigation plan must be developed that includes the entire mine life, from construction to operation and closure. Baseline study, environmental assessment, environmental impact statements, and remediation plans are required as part of the permitting process.

To the extent that it can be monetized, the cost of the environmental mitigation plan must be included in the cash-flow analysis.

Input variables that are likely to influence the environmental bottom line are listed below. The list will vary among projects. It is the responsibility of the environmental analysts to make sure that all appropriate variables are taken into account.

Water

The impact of mining operations on water is one of the most important contributors to the environmental bottom line. Water is also one of the most significant factors influencing the social and, to a variable extent, the financial bottom lines. Consideration must be given to the following factors:

- Water consumption. How much water will the mining and processing operations consume? How much potable water will be needed? Where will these waters come from? Will dams, water retention, and distribution structures be needed to satisfy operational needs? What will be the impact on current users, including humans, animals, and vegetal life?
- Disruption of surface water. Will pit design and surface infrastructure require relocation or disruption of surface water, including springs, wells, streams, rivers, irrigation channels, lakes, and swamps? Will mine dewatering impact surface waters? Were all consequences of these relocations or disruptions assessed and mitigation plans developed?
- Disruption of underground aquifers. Were the likely consequences of mine dewatering on shallow and deep aquifers taken into account? Will underground waters be used by mine, process, or drinking water and was the impact of this consumption evaluated?
- Acid mine drainage. Acid mine drainage and heavy metal contamination is the result of oxidation of sulfides in open-pit and underground mines, waste dumps, and stockpiles. If not controlled, it is likely to be a main source of water contamination. Were tests completed to determine the acid-generating potential of mined rock? Were appropriate facilities designed to control drainage from potentially acid-generating stockpiles and waste dumps, and to detect and control water contamination if it occurs?
- Pollution of surface water. What is the expected quality of water discharged from mine dewatering? Were the capital and operating costs of a water treatment plant included in the cash-flow analysis? Were sediment and erosion control plans designed and its cost included in the cash-flow analysis?

- Pollution of underground aquifers. Would mine dewatering, and water flows through open pits and underground openings result in aquifer contamination? What precautions were taken to avoid leakage of contaminated waters into aquifers from tailings, leach pads, and waste dumps? If in situ leaching is to be used, will it result in aquifer contamination? If injection wells are to be used for excess water, how will the water quality be monitored?
- Pollution of the oceans. Mines near or under the ocean have the potential to contaminate near-shore waters, thus affecting plants, animals, fisheries, and so forth. Underwater tailings disposal raises many unresolved questions. Tailings storage facilities and waste dumps have the potential to contaminate oceans if located near rivers or near the seashore.

Soil Contamination

- Were possible sources of soil contamination identified, including disposal of solid and hazardous waste, soil erosion and sedimentation, water or airborne contamination originating from pit operations, haulage roads, waste dumps, stockpiles, tailings, and leach pads?
- Was a mitigation plan developed to control or prevent surface erosion and soil contamination? Was the cost of such a plan included in the cash-flow analysis?

Air Pollution

- Were sources of air pollution, including pit operations, haulage roads, waste dumps, stockpiles, tailings, and leach pads, identified?
- Was consideration given to gas emissions from power generation, mineral processing operations, exhaust from mobile sources (open-pit and underground equipment, light vehicles, traffic from and to the mine)?
- Were appropriate plans developed to prevent, control, or mitigate air pollution, and their costs estimated?

Noise Pollution

- Were all sources identified, including those originating from mining, processing, and vehicular traffic?
- Was a mitigation plan designed and its cost estimated?

Visual Impact

Was the visual impact taken into account in locating and designing waste dumps, leach facilities, processing plant, offices, and other mine infrastructures?

Impact of Blast Vibrations and Mine Subsidence

Vibrations originating from mining and processing operations can affect the stability of mine infrastructures, buildings, and homes of people living nearby. Underground and in situ mining can result in surface subsidence and sinkholes. Excessive proximity of open pits to existing structures can result in instabilities. Were these considerations taken into account?

Effect of Mining Activities on Biological Environment

- Mining projects impact all forms of wildlife—aerial, terrestrial, and aquatic—through habitat loss, habitat fragmentation, destruction or reduction of feeding ground, water disruption, and contamination.
- Access roads, railroads, and other infrastructure influence animal migration.
- Deforestation, clearing of vegetation, and erosion have a direct impact on vegetal and animal life and biodiversity.
- Endangered or threatened species are protected and their presence can have a significant impact on project design, cost, and technical and financial feasibility.

Mine Closure

The ideal objective of mine closure is to return the site to conditions that most resemble pre-mining conditions through land rehabilitation and development of future land use. Reaching such an objective is rarely, if ever, technically or economically feasible. Sustainability is a more realistic objective, which requires that after-closure site utilization remains beneficial to future generations. Practical solutions must take into account rules and regulations, company values, and societal expectations, subject to constraints imposed by technical and financial feasibility. When evaluating mine closure, the environmental analyst should consider the following:

- The end-of-life site reclamation and closure plan must include open-pit and underground mines, processing facilities, waste dumps, tailings, leach pads, buildings, roads, and all other facilities that will no longer be used.
- Was a detailed assessment completed of the environmental and social conditions likely to prevail at the end of the mine life and beyond? Were impact mitigation and closure plans developed accordingly?
- Was consideration given to continuous reclamation throughout the life of the project to minimize environmental impact and cost of closure?
- Mine closure must take into consideration long-term impacts on surface and underground water flows and water quality, including those resulting from formation of pit lakes and flooding of underground workings. Will there be a need for water treatment in perpetuity?
- Will appropriate funds be set aside to ensure that the costs of reclamation and closure will be paid for by the project, including costs to perpetuity? Governments are increasingly regulating this process. Plans to pay for reclamation of one project using profits from the next project are becoming increasingly unacceptable. Such an approach to reclamation presents risks typical of a Ponzi scheme: Who is going to pay if the next project does not materialize or the mining company goes bankrupt?

Catastrophic Failures and Natural Disasters

The probability of catastrophic failures, resulting environmental and social impacts, and remediation costs must be fully assessed, including

- Failure of tailings dams, waste dumps, water retention dams, pit walls, underground openings;
- Overflow from ponds and dams;

- Failure of water treatment plants;
- Failure of processing plant environmental control facilities; and
- Extreme natural disasters, such as earthquakes, tsunamis, floods, droughts, forest or brush fires, landslides, avalanches, and extreme heat or cold weather.

The likelihood that catastrophic natural disasters will occur during the life of the project can be estimated, but the timing of such events is unknown. A 100-year flood is just as likely to happen in the next year or two as it is in 100 years. Climate change is increasing the likelihood and magnitude of catastrophic natural disasters. While the probability of such events may be very low, the consequences can be disastrous and should be taken into account when designing the project.

Uncertainty and Risk Assessment

The list of variables that can influence the environmental bottom line shows the difficulty that one may have in estimating the likelihood of their potential impact on the project, the cost of this impact, and whether or not this impact can be effectively mitigated. The impact of some variables can be reduced by changing mine, processing plant, and infrastructure location and design. Some impact prevention and remediation costs can be quantified in monetary terms and should be taken into account when estimating the financial bottom line. Other impacts are not quantifiable in monetary terms. Impacts that are not acceptable or will need remediation are becoming increasingly regulated by governments. Situations exist where company expectations need to be more demanding than those imposed by regulations. Following regulatory requirements may not be sufficient to satisfy the needs and expectations of some key stakeholders, such as the local population or international NGOs. Ignoring or failure to manage these expectations might result in significant disruption to mine development and operations.

VARIABLES INFLUENCING THE SOCIAL BOTTOM LINE

...the dust which is stirred and beaten up by digging penetrates into the windpipe and lungs, and produces difficulties in breathing, and

> *the disease which the Greeks call "asthma." If the dust has corrosive qualities, it eats away the lungs, and implants consumption in the body.... For their digging they [the miners] should make for themselves not only boots of rawhide, but gloves long enough to reach the elbow, and they should fasten loose veils over their faces.*
>
> Georgius Agricola, *De Re Metallica*, 1556

The definition of a project's social impact is often understood to be limited to the impact that is felt by local communities. Indeed, this is likely to be the main factor to be taken into account. A broader definition is used here, where the social bottom line is defined as a measure of the impact, positive or negative, that the project is likely to have on all people, including all stakeholders:

- Shareholders
- Banks and financial institutions
- Business analysts
- NGOs
- Employees and their families
- Customers and suppliers
- Local, regional, federal, and country governments
- Local communities and organizations
- Future generations

General Considerations

Stakeholders have different interests, priorities, and values and therefore different views considering a project's social impact and the weight that this impact should be given in estimating the project utility. These often-conflicting opinions and priorities must be considered when estimating the social bottom line.

Each country has laws, rules, and regulations that specify minimum requirements on how to estimate, report, and mitigate the social impact of a project. These minimum requirements must be satisfied before obtaining the necessary governmental permits to develop and operate the project. Nevertheless, it is important to recognize that the expectations of many stakeholders often exceed what laws require. Local communities might view the project as a source of significant disturbance to their way of life with no compensating benefit. They might prevent development or stop operations even if the company has obtained all governmental permits. Failure to take into account all stakeholder expectations may result in underestimation of the social bottom line and loss of the social license to operate.

Mining can bring economic and social benefits to communities by creating jobs, improving roads, building schools, upgrading medical facilities, and increasing the demand for goods and services. But it can also create social changes that can lead to social tension and conflicts, sometimes violent. Such conflicts can result from lack of sufficient consultation and community engagement. The community may not trust information given by the mining company concerning the project's expected impact, positive and negative. There are likely to be differing expectations of social and economic benefits, environmental concerns, and disputes over land use and economic compensation. The situation is exacerbated if the local communities believe that their own government does not have their interest at heart but favors the development of a new mine because of expected tax revenues. Uneven distribution of costs and benefits represents a major concern if social costs are local while tax benefits are retained by the central government.

Meeting all regulatory requirements and obtaining a legal license to operate is no longer sufficient. Local communities and the NGOs that are committed to defending the interests of these communities often require that more be done. Early, open, continuous, and well-intended communication between mining company and key stakeholders is a fundamental requisite for the project to be accepted and allowed to operate in a safe and reliable environment.

The social bottom line takes into account the current level of acceptance that the stakeholders have in the proposed project. It also considers the measures taken to ensure that this acceptance is improved and maintained during the life of the project. The possibility that, and reasons why, social dissatisfaction may occur over time should be determined, and plans should be developed to mitigate the consequences of such dissatisfaction.

Minimum Requirements

Reliable estimation of the social bottom line depends on the following:

- Competence of the analyst responsible for evaluation of the social bottom line; competence of persons supplying information needed by the analyst to estimate the social bottom line.
- Whether all variables that influence the social bottom line have been properly identified and estimated, including the uncertainty of these estimates.
- Whether the process used to estimate the social bottom line takes into account all relevant input variables and combines them in a way that fairly represents the company's objectives and values.
- Knowledge of applicable rules and regulations, including social, environmental, and permitting requirements; knowledge and understanding of the mining company's social objectives, values, and ethical expectations.
- Knowledge of all applicable taxes, royalties, and dues, and understanding of whom, among the stakeholders, will benefit from them. Particular attention should be given to whether taxes and dues will benefit local communities. It should not be assumed that government, which has many constituencies to take into account, will automatically use tax revenues to mitigate the local impact by improving infrastructure, medical services, public safety, and so on.
- The mining company should determine the extent to which it may have to improve infrastructure and supply services at its own cost, in areas that might otherwise be considered as the responsibility of government.
- Understanding of the social expectations of local communities and NGOs. These are likely to differ from and be more taxing than the governmental regulatory requirements and company expectations.

- A baseline survey of the communities that are likely to be impacted by the project should be completed as soon as there is evidence that an economic project is likely to be defined.
- An appropriate social impact mitigation plan should be developed that includes the entire mine life, from construction to operation and closure. Environmental assessment and environmental impact statements are required as part of the permitting process, which include social impact and a mitigation plan.
- To the extent that it can be monetized, the cost of the social impact mitigation plan must be included in the cash-flow analysis.

A challenge in evaluating the social bottom line is that it is a function of perceptions and expectations of human beings, which are not only difficult to determine at a given time but also likely to change over time. Even if the best communication plan has been developed and is continuously implemented and adapted to accommodate change, events might still take place that could significantly affect the value of the project. Examples of such events are strikes, blockage of access to the mine site, threats to employees and management, terrorism, and so on. Political risk resulting from change of government or of governmental rules and regulations fall into this category. Action plans should be developed to continuously monitor the social and governmental environment to determine the likelihood of such events, take preventive actions, and react appropriately if such events do occur.

Communication

The main difference between the environmental and social bottom lines is that the social bottom line relates to people—their beliefs, values, emotions, past experiences, expectations, levels of trust, fears, greed, priorities, attachment to family, tribe, community, religion and culture, and general psychological and sociological motivations. For these reasons, communication arguably represents the most critical element in the evaluation, control, and management of the social bottom line.

Minimum requirements for effective communication between mining company and stakeholders include the following:

- Knowledge of stock exchange and other regulatory requirements concerning public reporting of exploration results, mineral resources, mineral reserves, financial results, and forward-looking statements concerning projects.
- Knowledge of country-specific legal and regulatory requirements concerning environmental and social impacts, including baseline studies, impact assessment, and mitigation plan, and communication of this information with government and stakeholders.
- Although following rules and regulations may be sufficient to obtain governmental permits, expectations from other key stakeholders are likely to exceed government requirements. Understanding this difference is vital to gain and keep the social license to operate.
- Effective communications must take into account Internet and social networks that are used to disseminate both positive and negative information and opinions concerning the mining company and its operations.
- Communicating with local, regional, national, and international organizations, including NGOs, must be initiated early to give the company time to understand and manage diverse and changing expectations.
- Communication must be planned, open, transparent, and aimed at gaining trust and mutual understanding.
- Engagement with local communities must start with understanding who is locally responsible, both legally and factually, to make decisions and influence society.
- Respecting and listening to communities is critical to gain understanding of their concerns or fears of the unknown, to be aware of the rumors that may be circulating, and to preempt unrealistic expectations that will have to be managed.
- Understanding and respecting local culture, traditions, and religion is a requirement for effective communication.

- To be locally effective, communication is likely to require the ability to speak local native languages, such as Quechua or Mapudungun as well as Spanish in some parts of South America, Kiswahili in southeastern Africa, French in some parts of Canada and Africa, and so on.
- Local communities and local governments do not necessarily trust representatives of the central government and their ability or willingness to protect local interests. Companies need to be aware of such situations and develop processes to ensure both central and local acceptance.

Shareholders, Banks, and Financial Institutions

Shareholders, banks, and financial institutions are primarily interested in the financial bottom line, the risk to their investment if financial performance is less than expected, and whether the mining company has in place appropriate risk mitigation plans. Depending on risk tolerance profiles, some investors may be more attracted by the possibility of exceeding expectations. But downward risk, as perceived by the investor on the basis of information supplied by the company, is always a major concern. While ethical values may drive consideration of the probable environmental and social impacts of the project, the main concern is likely to be whether these impacts will compromise the public image of both the mining company and the stakeholders, and could result in the mining company losing its social license to operate, both locally and worldwide.

Public communication with shareholders is extensively regulated by stock exchanges, with regulations differing between countries. To satisfy expectations, mining companies may choose to publicly supply information that exceeds regulatory requirements.

Financial Analysts

A role of financial analysts is to use their knowledge of the mining industry, the company, and its projects to make recommendations to shareholders on whether to buy or sell company shares. To make fair recommendations, they need to have access to reliable information.

Financial analysts can have a significant influence, both positive and negative, on shareholders' opinion, share prices, and the public image of the company. Chief executive officers give significant attention to the financial analysts' public statements and act accordingly. Financial analysts do not always have the knowledge or experience to make sound recommendations, and giving excessive consideration to their opinion can result in decisions detrimental to the project, the company, and the shareholders. It is the responsibility of management to decide which opinion should be taken into account and to ensure that financial analysts have access to reliable information.

Nongovernmental Organizations

Many NGOs view their role to be protectors of the environment and defenders of the interests of local communities. The project manager should expect NGOs to take an early interest in possible environmental and social impacts. Because the reach of NGOs is international, their opinion can have international repercussions. Criticism of actions taken by the company in one location can compromise the company's social license to operate, not only where the project is located but also with other projects worldwide. The project objectives should be presented to the NGOs early in the process, and these presentations should emphasize how environmental and social concerns are taken into consideration. Some NGOs are systematically anti-mining, a position that is unrealistic, not sustainable, and may render any form of communication fruitless.

Employees and Their Families

The concerns of employees include job security, fair compensation, a healthy and safe working environment, and a reasonable balance between work and leisure. Depending on the project location, other considerations include

- Role of nationals and expatriates (These respective roles are often regulated by government with the percentage of expatriates expected to decrease over time.);
- Training of nationals and a plan for replacing expatriates with nationals;

- Salary and benefits accrued by expatriates and how they compare with those of local employees in similar positions;
- Expectations that the mining company is knowledgeable of and respects local culture, practices, and religions; and
- Appropriate housing, recreation facilities, food supply, quality of life, health services, safety and security, both at work and outside of work.

Understanding and managing employees' expectations is another important part of what is needed to estimate the social bottom line. Failure to do so can result in work slowdown, strikes (legal or illegal), and other more extreme reactions from employees.

Customers and Suppliers

The first concern that a mining company has when considering a new mine is whether there will be a market for its product (output) and whether appropriate supplies (input) can be obtained at a reasonable cost. The role of *customers*, defined as those who purchase the product of the mining operation, varies significantly depending on the mineral being mined. A gold mining company's objective might be to minimize the cost of sales with appropriate contracts with refineries. A company that produces industrial minerals will give particular attention to the quality of the product requested by each customer and the probability that this quality will be maintained when different parts of the deposit are mined. A company that sells copper concentrates, which will be assayed by both seller and buyer, must ensure that appropriate procedures are in place to facilitate reaching agreement on grade, quality, and dollar value of product sold. Beside formal contracts, open communication with customers is useful to create the level of trust and understanding needed to minimize the effect of production difficulties.

Suppliers of goods and services play a critical role during both construction and operation of a project. Early contacts are needed to ascertain the availability and cost of supplies. Once a project is sufficiently advanced, contracts are needed to reduce supply risks. The mining company should give appropriate consideration to developing and using suppliers from the region and country where the mining operation is located, thus improving social relations and fair distribution of benefits. Planning of inbound and outbound logistics is an important part of ensuring accurate estimation of capital and operating costs.

Local, Regional, Federal, or Country Governments

> *The miner should make careful and thorough investigation concerning the lord of the locality, whether he be a just and good man, or a tyrant, for the latter oppresses men by force of his authority, and seizes their possessions for himself; but the former governs justly and lawfully and serves the common good.*
>
> Georgius Agricola, *De Re Metallica*, 1556

All levels of government, local, regional, or national, can reasonably expect to benefit from a new mining operation. This benefit is in the form of taxes, royalties, employment, development of infrastructure, and increased level of economic activity. However, there can be a local perception that the negative impact of mining occurs locally while the positive impact in the form of tax revenues is centralized and not fairly distributed. Local communities might expect the mining company to be responsible for mitigating the local impact of the operation while the mining company might consider it the responsibility of government.

To assess the impact of the project, the company should evaluate the answers to the following questions:

- Will benefits derived from mining, including taxes, be distributed fairly between local, regional, and national governments?
- Will the need to improve local infrastructure (roads, electricity distribution, availability of clean water, sewage system, medical facilities, food distribution, police, and safety) be paid by taxes or will it require additional contributions from the company?
- Is unfair local distribution of benefits a consequence of corruption and bribes? The mining company must have clear rules on how to deal with corruption, bribes, and other types of potentially illegal

facilitating payments, especially in countries where such business practices are expected. Knowledge of national and international regulations concerning such business practices must be understood by all those involved in the project.

- Political risk, including risk of creeping or outright nationalization, cancellation of contracts, changes in tax structure and in other rules and regulations should be evaluated and included in evaluation of the risk to the social and financial bottom lines.

Local Communities

The most significant social impact of mining operations is likely to be on local communities. Although some negative impact may be unavoidable, the company must ensure that the positive side is dominant. Local communities have the power to significantly influence or even revoke the company's social license to operate. A project is not likely to succeed economically unless early and careful attention is given to the needs and expectations of these critical stakeholders. They might take matters into their own hands if they believe that their government does not adequately represent them and their interests are not sufficiently protected.

Government regulations require that the environmental and social impacts be assessed early and that environmental and social impact statements be submitted, including impact mitigation plans, before approval is granted for construction and operation. However, such requirements are not likely to fully account for expectations from local communities.

The positive impact will be greatest if taxes paid by the company are equitably distributed and if the company actively contributes to improving the quality of life of local residents through direct investment, focused charitable contributions, and other local activities. Positive contributions include

- Additions to the local economy through training, employment, salaries, local purchase of goods and services, and development of new business opportunities;
- Improved quality and access to health facilities;

- Improved opportunity for education (better schools);
- Infrastructure improvement (roads);
- Improved water supply and water quality (water treatment plant, public water distribution network, sewage disposal);
- Improved availability and reliability of electrical power distribution and other sources of energy;
- New recreational facilities; and
- Better opportunities for younger generations (education, quality of life, work opportunities).

Some of the negative impacts of mining, whose likelihood, magnitude, and possible mitigation need to be assessed to fully value the social bottom line, include the following:

- Land acquisition resulting in human displacement and resettlement (impact on landowners including loss of homes, land, and livelihood; loss of cultural and spiritual environment; and need to find new housing and occupation)
- Difficult adaptation of local communities from subsistence or agricultural economy to an unfamiliar industrial work and life environment
- Impact on availability and cost of housing, food, water and power supply resulting from new demand for mining operation
- Impact on surface and underground water quality and availability, including drinking and agricultural water
- Mine infrastructure interfering with travel and traditional access routes
- Pressure on local infrastructure, and increased traffic, noise, dust, and visual pollution
- Impact on public health, including physical and mental health and social well-being
- Social disturbances due to influx of immigrants looking for new job opportunities or supplying services initially not available locally

- Transient male workforce, especially during project construction, which increases alcohol and drug availability and consumption, prostitution, and crime (mitigation requires prevention, education, improved security, and medical facilities)
- Disruption of traditional economic interactions and existing social status hierarchies

Future Generations

Sustainability demands that not only present- and medium-term impacts are taken into consideration but also impacts on future generations and the environment that are likely to continue after the operation is closed. Emission of greenhouse gases that may contribute to global warming and climate change should be considered when designing the operation. The closure plan must assess the environment, both social and natural, that will remain after the mine shuts down. Beneficial postmining land use must be considered, as well as the need to mitigate long-term impacts such as those resulting from mine flooding after termination of mine dewatering activities.

Catastrophic Failures and Natural Disasters

In addition to risk related to changes in human expectations, the risk of mine-related disasters can also have very significant social impacts. Included are failures of water dams, tailings impoundments, waste dumps, heap leach facilities, and emission control equipment, which can result in contamination of soil, water, or air and a significant impact on employees, communities, and the natural environment. When designing and locating infrastructure, offices, housing, relocation sites, and so forth, one should take into account the likelihood of such catastrophic events, and mitigation plans should be developed in case such events occur.

CHAPTER 12

GEOLOGY AND DEPOSIT CHARACTERIZATION

> *By tests miners can determine with certainty whether ores contain any metal in them or not; or if it has already been indicated that the ore contains one or more metals, the tests show whether it is much or little; the miners also ascertain by such tests the method by which the metal can be separated from that part of the ore devoid of it.... Unless these tests have been carefully applied before the metals are melted out, the ore cannot be smelted without great loss to the owners, for the parts which do not easily melt in the fire carry the metals off with them or consume them.*
>
> Georgius Agricola, *De Re Metallica*, 1556

In Chapter 11 variables were analyzed that influence the three bottom lines, financial, environmental, and social, that define the utility of a project. The properties of these variables are primarily defined by decisions made by the technical departments within the mining company (Figure 10.4). Sound assessment of project risk requires understanding the role that each technical department must play, the emphasis being on the information that the department is responsible for gathering and analyzing, and how this information contributes directly or indirectly to the value of the three bottom lines. The role of geology is the subject of this chapter. The other technical departments are analyzed in subsequent chapters.

GENERAL CONSIDERATIONS

Geology is defined here as inclusive of all the earth properties that need to be characterized to develop safe, sustainable mining operations. It therefore includes deposit geology (structure, mineralogy, alteration, lithology, grade) as well as geometallurgical, geochemical, and geotechnical ore and waste characterization, hydrology, and soil and subsoil descriptions. Geological properties are quintessential space-related risk variables, as they can be observed at specific locations, such as down drill holes, but must be inferred between observation points.

Geology is at the beginning of the data-gathering process. Information flows from the geology department to the mining, processing, infrastructure, and eventually financial analysis departments. The primary concern for the geology department should be to ensure that the downstream needs for information are well understood and satisfied. Conversely, the departments that need geologic information to make their own decisions must clearly define their upstream requirements as well as the information that they must generate to satisfy the needs of the next downstream departments.

The geology department is responsible for the discovery of new deposits and characterization of these deposits once they are discovered. The following tasks must be performed to ensure sound decisions:

- Determine which geological information is needed to evaluate a deposit. This requires effective communication with mining, processing, and environmental departments, as well as understanding of likely infrastructure requirements.
- Perform those activities that are needed to get sufficient, representative, and reliable data. The geological properties of mineral deposits that impact project evaluation include location, geological setting, deposit type, style of mineralization, controls of mineralization, grade of mineralized material, lithology, mineralogy, alteration, material type, bulk density, structures; geochemical, geotechnical, hydrogeological, and geometallurgical properties; and soil and waste characterization.

- Interpret data and develop three-dimensional (3-D) models on the basis of the acquired information. Geologic information is rarely used in its raw form. Mining engineers, metallurgists, process engineers, environmentalists, and infrastructure engineers need 3-D models to make critical decisions and to fully evaluate the economic feasibility of a project. To ensure relevance and completeness, communication with those who will use the models should be initiated before model development.
- Transfer the model to the users, making sure that the limitations of the model are well understood. Review of the model with the users may indicate that additional data are needed to ensure completeness, improve reliability, and manage project risk resulting from geologic uncertainty.

The project risk is directly related to the level of understanding of the geological properties of the deposit. Geology determines the quantity and quality of material available in the ground. Mining methods that are technically feasible depend on geology, and so are the methods that can be used to process the material that will be mined and to obtain a salable product. The design of safe open-pit and underground mining operations requires sound geologic information. Geology and presence of deleterious elements influence the design of environmentally sound processes. Geology plays an important role in sizing both mine and processing plant, and in locating surface and underground facilities and infrastructure. The environmental characteristics of waste dumps, stockpiles, and tailings dams must be determined by gathering geological information. Mine dewatering, water availability, and water quality are dependent on geology. Ultimately, the capital and operating costs of a new mine are to a large extent defined by geology.

RELATIONSHIP BETWEEN GEOLOGY, MINING, AND PROCESSING

If we make the simple assumption of a linear decision-making process, the relationship between geology, resource modeling, mining, and processing can be represented by Figure 12.1. The underlying assumption is that the only input necessary to develop a mine plan is a 3-D model of the quality/grade of

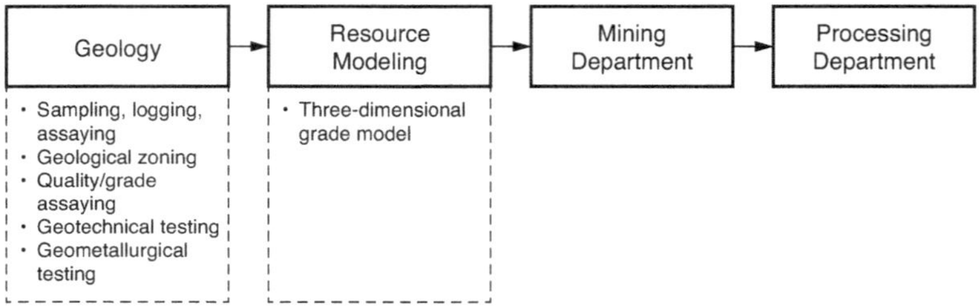

FIGURE 12.1 Linear (oversimplified) process to define geologic and resource modeling requirements

material available to be mined. Such a simple assumption may be true in rare situations where knowing the quantity and quality of mineralized material in the ground is sufficient to justify a new mine. Evaluation of near-surface, highly oxidized, dump-leachable gold mineralization is arguably an example of a situation that most closely resembles the Figure 12.1 model. However, even then, companies are cognizant of the need to make sure that the material can be mined and processed safely, that a salable product can be obtained, and that revenues from sales will exceed capital and operating costs. Geologic mapping is required, a grade model must be developed, and metallurgical tests must be completed before a determination can be made that there is the potential for eventual economic extraction.

In actuality, the exploration process and the information that this process is expected to generate are significantly more complex. A realistic process representation may look like that shown in Figure 12.2. This figure illustrates the wide variety of information that must be available to develop mine plans and production schedules and to estimate capital and operating costs. Tonnages and grades of mineralized material remain of primary concern. If there is not sufficient mineralized material of high enough quality, there is no mine. But considerably more information is needed to be able to ascertain that indeed there is the potential for economic extraction. The spatial distribution of the mineralization and how tonnage, grade, and continuity vary in function of cutoff grade represent critical information required by mining engineers. The quality of the material that can be delivered to the processing plant must be defined. The relationship between mine and mill performance, and material type must be determined. Waste material must be characterized according to its environmental properties. This information is needed to decide on the

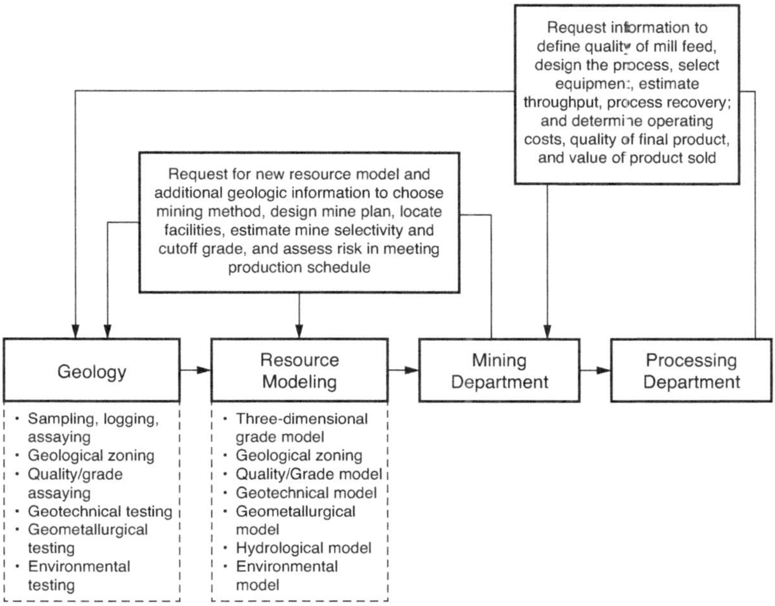

FIGURE 12.2 Complex (realistic) process to define geologic and resource modeling requirements

best mining method, optimize cutoff grades, design mine plans, develop production schedules, determine infrastructure requirements, and estimate corresponding capital and operating costs and cash flows.

DEPOSIT CHARACTERIZATION AS A MULTISTAGE PROCESS

Exploration projects are divided into several stages. The decision to go from one stage to the next is a function of the results obtained. The exploration process was discussed in Chapter 7 where use of decision trees was suggested as a means to illustrate objectives, quantify probabilities, and analyze options. Justification of an exploration stage requires specification of the following.

- Objective. What is this stage expected to demonstrate? Which criteria will be used to determine whether success is achieved?
- Technical component. What will be completed during this stage? In early stages, this may include surface geological, geophysical, and geochemical surveys. In later stages, activities such as drilling,

sampling, core logging, assaying, and deposit modeling will be appropriate activities.

- Cost of completing this exploration stage. Time and monetary cost should be estimated, including financial, social, and environmental impacts.
- Probability of success and failure. What is the likelihood that the exploration stage, as technically defined, will reach the stated objectives?
- What the next stage will be if results obtained in the current stage are positive. A conceptual design of the next stage should be completed, including objective and cost.
- What the next stage will be if results obtained in the current stage are negative. Decisions may include refocusing the current stage (get more information or improve analysis of available information), postponing the project until conditions improve, selling it, or closing and abandoning the project.

In this chapter, exploration projects are assumed to be composed of two distinct stages: early stage and late stage. This assumption is a simplification of actual exploration stages. However, it is useful to illustrate the various types of risks that are faced when attempting to discover and characterize mineral deposits.

EARLY-STAGE EXPLORATION

The objective of early-stage exploration is to discover a mineralization that shows a potential for eventual economic extraction. To justify an exploration program, the geologist must estimate not only the cost of this program but also (1) the probability that it will be successful and (2) the value or range of values of the targets that the program is aimed at discovering. The probability of success is low, while the probability of failure is high. The cost of early-stage exploration will be incurred whether a deposit is discovered or not. The possible value of a discovery is uncertain. There are complex relationships among the exploration activities that are proposed, the value of the deposits that might be discovered, and the probability of discoveries.

The main factors that influence the probability of success include the competence of the geologists, the geologic potential of the area of interest, and the design and implementation of the exploration program. Examples of multistage exploration programs were given in Chapters 7 and 9. These examples showed how statistics, decision trees, and the utility function could help better manage exploration programs. It was shown that a well-designed exploration program could minimize costs while maximizing the probability of success. While somewhat oversimplified, Figure 12.1 can be viewed as an acceptable representation of the context within which the geology department is likely to operate during early-stage exploration. There is insufficient information to justify the complex feedbacks shown in Figure 12.2.

The cost of an exploration program must take into account the three bottom lines, financial, environmental, and social. The direct financial impact includes the costs of acquiring mineral rights; getting access to the property; geological, geophysical, and geochemical surveys; soil and stream sampling; and any planned subsurface sampling by drilling, trenching, or other means.

The environmental impact is mostly the result of surface disturbance resulting from accessing the exploration site and completing the exploration program. If drilling is part of the exploration program, the risk for contamination of underground aquifers must be assessed. Drilling is often subcontracted, but this does not reduce the responsibility of the mining company for actions taken by subcontractors. The environmental costs must include those that will be incurred for site remediation if exploration is not successful or if additional work is significantly delayed. Appropriate reclamation should be planned in advance, taking into account rehabilitation of all disturbed surfaces, including exploration camp, access roads, drill sites, and other workings. Particular attention should be given to ground contamination from drilling fluids, disposal of oil and other chemicals, and soil erosion, and to the need for appropriate closure of drill holes.

The impact of early exploration on the social bottom line can be significant and should not be underestimated. Early exploration is the first time that the mining company comes into contact with the local communities. The reaction of the communities to this new and unfamiliar intrusion into their traditional way of life is likely to govern their reaction to any future activity in the area. Ignoring the social impact of early exploration can delay, prevent, or increase the cost of any follow-up activities in the region, even if the results of the exploration program are positive. Actions to be taken to minimize the social impact of an early exploration program, not only for the duration of this program but more importantly to facilitate future activities, include the following:

- Early contact and formal discussions with local government and other organizations, civil or religious, which are respected in the local communities.
- Direct contact with the local population. This should take into account existing social structures. The ability to use the local language, which may not be English, shows respect for local culture and improves relationships.
- Open discussion of what the project is about, its expected impact, and remediation that will take place.
- Attention should be given to not creating unrealistic expectations, especially when the likelihood of future investments in the region, which depends on successful exploration, is low.
- Serious consideration should be given to hiring temporary help or using local services for camp construction, food supply, and procurement of smaller items such as core boxes, road maintenance, and remediation.

FINANCIAL EVALUATION OF EARLY-STAGE EXPLORATION: AN EXAMPLE

A simple example follows that illustrates the relationship between the cost of early-stage exploration, the probability of finding a deposit of economic value, and the value of such a deposit if it is discovered.

The project geologist has estimated the type of deposit that could be discovered in the area of interest. If such a deposit is present and discovered, the net present value (NPV) of the cash flow that could be generated is of the order of \$1 billion:

$$\text{NPV} = \$1 \text{ billion}$$

The geologist estimates that the probability that there is such a deposit in the area being explored is $p_1 = 5\%$, and the probability that it will be discovered is $p_2 = 60\%$. In other words, the probability that the exploration program will be successful is

$$p = p_1 \times p_2 = 5\% \times 60\% = 3\%$$

The probability p_1 is a function of the geology of the area being explored and the characteristics (represented by NPV) of the exploration target. The probability p_2 is a function of the nature of the exploration program (represented by its cost C) and the target characteristics.

Excluding exploration costs, the expected value of the exploration program is

$$p \times NPV = M\$30$$

The estimated cost C of the proposed exploration program must include acquisition of necessary permits, land access, construction of the exploration camp, completion of the exploration program (which may include surface activities, drilling, sampling, testing, data analysis and logistics), and site reclamation when the program is completed.

The cost C will be incurred whether exploration is successful or not. To be justified, this cost should be less than the expected value of a discovery weighted by the probability of discovery:

$$C < p \times NPV$$

The cost of an exploration program is typically very low when compared to the cost of developing a new mine. If C = M\$10 and the capital cost of developing a new mine with a \$1-billion NPV is expected to be I = M\$2,000[1], the ratio C/I is

$$C/I = M\$10/M\$2{,}000 = 0.5\%$$

This very low cost ratio is not a significant factor in determining whether a first-stage exploration program is justified. The only significant parameters are p and NPV. The same statement cannot be made during later-stage exploration, when a deposit has been discovered and detailed delineation is

1 The capital cost I (\$2 billion) is included in the project NPV (\$1 billion). The expected return on investment is ROI = NPV/I = 50%. At this early stage, there is only p = 3% probability that exploration will be successful and that such a return will be achieved.

needed. The cost of deposit characterization can be considerable, even if still significantly less than that of developing a new mine.

The equation $C < p \times NPV$ can also be used to determine the minimum NPV that an as-yet undiscovered deposit must have to justify the cost C of acquiring mineral rights to a given property and exploring it. To answer this question, the geologist must estimate the probability of success p. The minimum NPV must be greater than C/p. If \$1 billion is spent to acquire the mineral rights to a project and the probability of finding significant mineralization is one in four, the minimum NPV that a discovery must have is \$4 billion. If the probability is one in ten, the minimum NPV is \$10 billion.

In early-stage exploration, estimates of p and NPV are highly subjective. They must take into account prior knowledge of the geology of the area being explored. Research should be completed to determine whether areas with similar geology were previously explored in which significant mineralization was discovered and to establish the value of these discoveries. The geologist's experience, competence, and integrity are essential if meaningful decisions are to be made.

In this example, a simplifying assumption was made that the expected NPV of a discovery was the most important decision criterion, thus ignoring those costs of a social and environmental nature, which may not be quantifiable in dollar terms or may not be given appropriate attention. If properly managed, the impact on the environmental bottom line is likely to be small. The impact on the social bottom line can be highly significant if not recognized early in the planning process and acted upon before initiating on-the-ground activities.

These conclusions only apply to early-stage exploration, when the probability of failure is high and the amount of activity taking place on the ground is limited. The same cannot be said about later-stage exploration, when the probability of success is rapidly increasing, significantly larger amounts of geologic information are needed, and intensive drilling is required to get this information and fully define the deposit.

LATE-STAGE EXPLORATION AND DEPOSIT CHARACTERIZATION

Late-stage exploration is defined here as including all activities that are related to the geological characterization of a deposit that has been discovered and which presents a significant potential for eventual economic extraction. A mineral resource has been defined. The remaining question is whether all or part of this resource can be converted to a mineral reserve. The role of late-stage exploration is to get detailed geologic information about the deposit as needed to control the risk associated with such critical activities as mine design and production scheduling, development of a suitable metallurgical process, infrastructure design, estimation of capital and operating costs, and cash-flow analysis. Estimates of the three bottom lines, and the decisions made on the basis of these estimates, are only as reliable as the geologic information that is the foundation on which these estimates are built.

The financial, environmental, and social impacts of late-stage exploration are significant. The level of activity taking place on the ground, mostly in the form of drilling (including access roads, drilling platforms, disposition of drilling fluids, surface and underground water disturbance, sampling and sample preparation, security, etc.), is high. Off-site activities (including logistics, laboratory testing, data analyses, etc.) are substantial. The impact on the three bottom lines is likely to be important. For advanced projects, it may be possible to consider this impact as part of the much broader impact that the mining project will have when mine development takes place. However, the results of detailed exploration might be negative, and site reclamation might have to take place without the benefit of a mining operation to pay for it. Whether or not late-stage exploration can be capitalized or must be expensed depends on applicable accounting rules and whether a reserve is declared.

Information obtained during late-stage exploration influences decisions concerning all aspects of a project, including the following:

- Mining engineering. To choose a mining method, develop a mine plan and a production schedule, and determine costs, engineers require a 3-D model of the mineral deposit (grade, density, rock types, geometallurgical properties, etc.) as well as geotechnical and hydrological information and waste characterization.

METALLURGICAL TESTING AND SAMPLE SELECTION

Getting representative samples for metallurgical testing presents specific challenges. Three options are generally considered, and each presents its own advantages and disadvantages.

- **Option 1.** Select a large number of small samples (a few kilograms each) throughout the deposit. The samples should be taken from existing core holes and tested individually. The tests can be run in laboratories using relatively inexpensive procedures such as bottle roll tests or tests using bench-scale flotation cells. The samples are chosen to give reasonable spatial representation of the deposit properties as needed to develop a three-dimensional geometallurgical model. The model is used in mine planning to characterize the material to be delivered to the processing plant over time and to design the process. On the downside, the properties of small samples are likely to differ significantly from those of thousands of metric tons of material to be processed daily in a full-scale commercial plant. Considerable progress has been made on how to use laboratory test results to predict commercial plant performance. Extrapolation from small scale to large scale adds a level of uncertainty that must be taken into account in plant design and when assessing the project risk.
- **Option 2.** Metallurgical tests can be completed on larger samples weighing 1 metric ton or more. For example, instead of using bottles to test the cyanide leachability of gold ore, one can use columns of increasing sizes. The samples used in these tests are often prepared by combining core samples to obtain material whose properties might be closer to those of expected mill feed. In some cases, large-diameter core holes may be specifically drilled to obtain material whose physical properties can be used to improve design of the comminution circuit. The number of samples that are tested is likely to be an order of magnitude smaller than that in the previous example. When using large-diameter drill holes, only a few areas of the deposit can be tested. Tests are limited to a small number of large geological domains.

Sidebar continues next page

- Process engineering. To design a suitable processing plant, engineers need quantity and quality of material to be delivered by the mine, including information concerning mineralogy, hardness, grade of salable products, and contaminants.

METALLURGICAL TESTING AND SAMPLE SELECTION

- **Option 3.** Even larger metallurgical tests can be completed in pilot plants designed to process a few hundred metric tons of material. Such large tonnages are most easily obtained from underground exploration drifts. This source of material presents the advantage of supplying coarse material that was blasted under conditions that can be similar to those that will prevail during mine operations. However, exploration drifts are not likely to give access to deep material or to a large variety of geological domains. The test results are indicative of what may be found in that part of the deposit that was accessible at the time the test was run. This may or may not be representative of other ore types or deeper material.

This discussion illustrates the complexities that the geologist must take into account when deciding which samples to make available to the metallurgist. The decision must not be made in isolation. Geologist and metallurgist must consult each other to ensure that there is common understanding of the test objectives as well as the strengths and weaknesses of each method.

- Environmental engineering. To assess and minimize the potential environmental impact, engineers require information concerning acid-generating and acid-neutralizing potential of waste material, thickness and characteristics of topsoil, bedrock properties of areas where facilities will be located, and both surface and underground hydrological information.
- Infrastructure engineering. To locate and design project infrastructure, including the processing plant, tailings storage facilities, waste dumps, leach pads, water dams, offices, laboratories, access roads, pipelines, and so on, engineers need both surface and underground geologic, geotechnical, and hydrological information.
- Marketing. The profitability of a mining project depends on the quantity and quality of marketable product that the processing plant will deliver to market. Product quantity and quality are a function of process design and the quantity and quality of plant feed.

An important part of late-stage deposit characterization consists of developing appropriate 3-D models. This is discussed in Chapter 13. Information needed to make well-informed decisions concerning mining, processing, and infrastructure can be found in Chapters 14, 15, and 16, respectively.

COMMUNICATIONS BETWEEN GEOLOGIST AND METALLURGIST

When communications between the geologist, mining engineer, and metallurgist fail, wrong and costly decisions are made. In this example, a deposit was being evaluated that contained high-sulfide gold mineralization. Laboratory tests had been completed using core samples, from which it was determined that ore roasting followed by cyanide leaching would be the most efficient process. The metallurgists recommended that a large-scale pilot-plant test be completed to improve process design and determine the need for environmental control of gas emissions. A minimum 200 metric tons of high-grade material was requested from the geologist. The objective was to improve the design of the processing plant and characterize the expected roaster's gas emission. Exploration drifts gave easy access to high-grade material, which was mined, placed in sealed barrels, and shipped to the pilot plant. The test results were used to develop a detailed process flow sheet, including roaster, cyanide leach plant, and environmental control facilities.

A project review meeting was scheduled to discuss the results of the pilot-plant tests. Project manager, geologist, mining engineer, metallurgist, and all other members of the project team participated. During this meeting, the following issues were quickly realized:

- The material used for testing represented only a small portion of the deposit.
- For the project to be economically feasible, a relatively low cutoff grade would have to be used. The average grade of the material to be processed would likely be half that of the material tested in the pilot plant.
- The content of deleterious elements in the high-grade material that was tested was much lower than that in the lower-grade material that would have to be processed.
- The roaster design and environmental controls would not be suitable to treat the available lower-grade material.
- New representative material would have to be selected and tested, and new designs would have to be developed.

What went wrong? The answer is quite evident: "high grade" did not mean the same thing to the geologist and the process engineer. To the geologist, it was the higher-grade material expected to be encountered in the mine. To the metallurgist, it was the average grade of material to be fed to the roaster. Communication between geological and metallurgical experts had failed because of different interpretation of the words *high grade*.

RISK MANAGEMENT AND DEPOSIT CHARACTERIZATION

Exploration is aimed at supplying information needed to reliably evaluate the utility of a potential mining project. To reach this objective, the following conditions must be satisfied:

- The project geologist must be fully qualified, having extensive and relevant experience in the managerial and technical aspects of deposit characterization and resource estimation. Experience must be with the type of deposit under consideration. He or she must understand what it takes to advance a mineralization from a geologic anomaly to a mineral deposit with potential economic value.
- Early and effective communication is necessary among the project geologist, mining engineer, and metallurgist. Such communication should take place as soon as a mineral resource can be inferred, with recognition that mining and processing methods likely influence geologic data requirements.
- There must be a clear understanding of what geologic information will be needed as exploration progresses. Those who need geologic information to make their own decisions—mostly mining, metallurgical, environmental, and infrastructure engineers—must have sufficient expertise to recognize what information they will require. Those who supply this information—usually geologists, geotechnical engineers, and hydrologists—must be aware of these requirements and understand how to best satisfy them, including how to get the information and how to process and present it in a useful format to the end users.
- The geologic information that is needed must be spatially representative of the potential economic portion of the deposit, taking into account how this information will influence both short-term and long-term operations. The location of surface facilities and underground accesses is influenced by the expected final location of the mine. Detailed information is needed about areas to be mined during the first one to three years of the mine life. Less information is likely to be sufficient for later periods in the life-of-mine plan.

QUALITY ASSURANCE AND QUALITY CONTROL

This example illustrates why putting in place an apparently sophisticated laboratory QA/QC program does not guaranty quality. The program was aimed at ensuring the quality of assays obtained from drill-hole samples. The samples were split, crushed, and pulverized by the mining company, and the pulp was sent to an external laboratory for assaying. The QA/QC process consisted of sending pulp duplicates, standards, and blanks to the laboratory, as well as pulp splits to two other laboratories as part of a round-robin program. The mining company had installed sophisticated computer systems recommended by a world-renowned sampling expert to securely transfer data from the laboratory to a geological database, and to extract the results from the database on a monthly basis and complete appropriate statistical analyses.

An independent review of the QA/QC procedures that were followed on-site showed the following:

- The computer-generated results of the statistical analyses were carefully filed but never reviewed. The junior geologist in charge of QA/QC did not have the expertise or receive the training necessary to analyze the results and make recommendations.
- The QA/QC program only considered errors that could be attributed to the assaying laboratory. Provided a reputable assay laboratory is used, these errors are likely to be small compared with those occurring during sampling, sample splitting, crushing, pulverizing, and preparation of the pulp samples sent to the assay laboratory.

Sidebar continues next page

- Processes must be developed to ensure that the information being gathered is sufficient and reliable. Information is reliable if it is both accurate and precise. Such processes must be effective without being bureaucratic or overly expensive. Quality assurance and quality control (QA/QC) procedures, training, and appropriate supervision are necessary throughout the data collection process, including surveying, drilling, sampling, core logging, sample preparation, testing, data reporting, storage, and security. Written procedures are needed, but procedures are of no value if they are not followed.

QUALITY ASSURANCE AND QUALITY CONTROL

- Errors occurring during sampling, splitting, crushing, pulverizing, and subsampling were not quantified.
- Visual inspection of the mining company practices indicated that significant errors were likely to be introduced during sampling and sample preparation. Quality assurance tests were completed, which confirmed that these errors were highly significant and were caused by loss of high-grade fines and biased splitting and subsampling procedures

The sample grades had been underestimated for approximately two years. Underestimation was highest in high-grade areas. Both tonnage and grade of minable material had been significantly underestimated.

In another mine, the grade of the material being processed was found to be significantly lower than expected. Many possible causes were considered, including errors in sampling, assaying, and modeling. Ultimately, it was found that one of the laboratories used during exploration was significantly biased upward. An expensive resampling and re-assaying program had to be implemented, followed by downward revision of publicly disclosed ore reserves and significant changes to the mine plan. This could have been avoided if QA/QC procedures had been followed at the time exploration took place.

- Pressure to get information quickly can result in the quality of the information being compromised. Poor-quality information will result in poor decisions or will have to be replaced with new information at a later date, thus increasing costs and creating delays.
- Contractors may be used for drilling, core logging, sample preparation, assaying, and metallurgical testing. Care should be taken in choosing and supervising these contractors. The mining company should be concerned not only with the quality of the work performed by the contractors but also by their respect of safety rules and communication with local communities.
- Whether exploration is completed in-house or is contracted to others, QA/QC should be used to randomly check the reliability of

MANAGEMENT MOTIVATORS AND RISK CONTROL

Case 1

There are two parts to risk: value and probability. Both must be considered when evaluating a resource. Company managers routinely emphasize the need to maximize net present value (NPV) and minimize delays to project development. Not enough attention may be given to the risk resulting from making decisions on early, insufficient, and possibly unreliable information. Overemphasizing the need to act quickly can motivate the project team to make premature positive recommendations concerning project development. The decrease in NPV resulting from project delays is easily quantified. Conversely, the probability and future cost of making the wrong decision are difficult to quantify and are often subjectively estimated. Unfortunately, the easy path is not necessarily the right one.

Case 2

Business analysts and exploration managers are acutely aware of the fact that exploration results directly influence share prices. Geologists are also aware of the fact that financing an exploration project can be greatly facilitated if drill-hole results are positive. These two motivators can result in priority being given to drilling areas of the deposit that are known to be high grade, even if these areas are too deep to influence expected project economics. Conversely, project risk reduction and NPV optimization would require geological characterization of lower-grade and marginal mineralization. In such situations, funds are allocated to satisfy short-term objectives (increase share price, ensure financing) while ignoring arguably more meaningful longer-term objectives (develop a financially successful mine operation). Management must remain cognizant of the motivators that guide exploration decisions and ensure that funds are allocated to optimize the company's objectives. In extreme cases, wrong motivators can result in misallocation of funds and shareholders receiving misleading information.

the results, including site visits, use of standards, duplicates, blank samples, and round-robin testing to compare laboratories.

- Commercially available databases are designed to store geological information. Databases must be reliable and secure. Information must be continuously backed up off-site to avoid losses resulting from total or partial corruption or destruction of the main database. Access to the database must be controlled to ensure that only legitimate additions, deletions, and accesses take place. A balance must be found between protecting information and allowing legitimate access.

- Databases must be regularly and randomly audited.
- The geology department is responsible for getting the information needed to make reliable mine planning, process design, and infrastructure decisions. It is the geologist's responsibility to ensure that representative samples are taken from the deposit on which appropriate tests can be completed. What is a representative sample depends on how test results will influence decisions.
- Decisions are usually not made on raw data. The geology department must interpret the information obtained from sampling and testing. Three-dimensional models must be developed which must be consistent with the geologic data. These models will be used by engineers to develop methods and processes, production schedules, cost estimates, and cash flows.
- The reliability of the models will depend on data quality, geologic interpretation, modeling method, and competence of the deposit modeler. This is further discussed in Chapter 13.
- For all geologic information, the chain of custody must be defined and documented. This is not only good practice to ensure quality and security but can also be a regulatory requirement.

EARLY-STAGE AND LATE-STAGE GEOLOGISTS

There can be significant psychological differences between geologists who are committed to the discovery of new deposits (early-stage exploration) and those whose responsibility it is to characterize deposits for eventual development (late-stage exploration). In early-stage exploration, the probability of success—making a discovery—is very low, but the rewards for success are very high. Learning from failure is important, but being confident that the next discovery is just around the corner is critical. In late-stage exploration—deposit delineation—the emphasis must be on a systematic approach to data collection, effective communication, quality control, and development of models that can be trusted by engineers to make sound decisions. Being realistic is the most important trait that should characterize late-stage geologists.

CHAPTER 13

RESOURCE MODELING

The responsibilities of the geology department include not only data collection and data analysis but also development of three-dimensional (3-D) models that are based on these data. The uncertainty concerning *model reliability*, defined as the extent to which the model represents the true state of nature, must be quantified and communicated to those who will use the model. Reliability depends on both data quality and modeling process.

GENERAL CONSIDERATIONS

The models that are needed vary with each project, depending on the decisions to be made, the geological complexity of the deposit and surrounding environment, the type of mineralization, and the expected mining and processing methods. At the beginning of a project, little information is likely to be available, but sound decisions concerning additional data requirements can be made using simple models. In later stages, when the project progresses from scoping study to pre-feasibility and full feasibility, increasingly detailed models are needed.

To assess model requirements, close and continuous consultations are needed between geologists, who are responsible for data acquisition and modeling, and those that need the models to determine applicable mining and processing methods, develop mine plans and production schedules, assess environmental impact, design supporting facilities and infrastructure, and develop capital costs, operating costs, and cash flows.

During early-stage exploration, when no or little drill-hole information is available, no 3-D model can be developed. However, the geologist should be able to describe the target that exploration is designed to discover. As exploration progresses, sufficient information may be available to estimate an inferred mineral resource (Chapter 4). At this stage, the only model that may be required is that of the mineralized part of the deposit. Even then, development of a meaningful model requires a preliminary understanding of the controls of mineralization, the constraints that mining will impose on selectivity, and the likely processing method.

When a project progresses from scoping study to pre-feasibility or feasibility study, the type and quality of information that is needed to estimate an indicated or measured mineral resource increases exponentially. A wide variety of models must be developed, including geology, grade, geometallurgy, hydrology, and so forth. Different levels of expertise are needed to develop each model. There are significant overlaps between model requirements, the common factor being geology. Compatibility between models should be an obvious requirement, even if often not maintained by inexperienced modelers. All modelers should have not only an in-depth understanding of the theoretical and practical aspects of the modeling methods that are applicable to the project and for which they are directly responsible, but also a general understanding of the entire project evaluation process, the many interrelationships between models, and how model uncertainty relates to project risk.

National and international codes for public reporting of exploration results, mineral resources, and mineral reserves require that the models used to evaluate a deposit be developed by, or under the responsibility of, a competent person (Chapter 4). Because of the significant difference in expertise needed to develop the various models, a number of competent persons must contribute under the guidance of a project manager. The project manager must accept responsibility for the quality of the work that is completed by the project team and the competent person.

Resource models are developed using computerized systems, which over the years have made the process easier, faster, and less subjective. The better systems are interactive, allowing the deposit modeler to modify computer-generated interpretations to take into account personal knowledge and experience. Geostatistics, a branch of statistics originally developed in the 1950s to improve estimation of the Witwatersrand gold grades, is widely accepted as the method to be used to estimate 3-D value distributions in all types of environments. The geostatistical methods that are currently available are many, based on a variety of theoretical assumptions and aimed at answering a variety of questions. To be considered a competent person the professional who accepts responsibility for the resource model should know the difference between different modeling methods, be able to determine which one is applicable in the situation under consideration, and have an understanding, both theoretical and practical, of the strengths and weaknesses that these methods present. (Competence implies understanding of which tools are available for a given job, how to use these tools, and what are the conditions under which the tools will give reliable results.)

While the amount of information that is needed varies with the project to be evaluated, it is likely that 3-D models will ultimately have to be developed to represent the following:

- Topography
- Geological characterization and zoning
- Three-dimensional grade distribution
- Geometallurgical properties
- Waste rock characterization
- Geotechnical properties
- Hydrology
- Soil and subsoil properties

A discussion follows of the various models that may be needed to fully evaluate a project.

TOPOGRAPHIC MODEL

A topographic model is needed for mine planning and to locate and design infrastructure, including processing plant, maintenance shops, laboratories, offices, housing, roads, pipelines, waste dumps, tailings storage facilities, water reservoirs, leach pads, and so on. Initially the topographic models are the responsibility of the geology department, which is the only one represented on-site. These models tend to cover a relatively small area. When the project advances, the area to be covered increases substantially and the

TOPOGRAPHIC MODEL AND INFRASTRUCTURE DESIGN

Two mineral deposits were discovered approximately 5 km apart. The processing plant was to be located near the larger deposit with material from the smaller one being transported to the plant by conveyor. A consultant was contracted to design and cost the conveyor.

Two separate topographic maps were available from the geology department, one for each deposit, with only limited information for the area separating the deposits. The consultant merged the two files and designed the conveyor accordingly. When the project was reviewed by the mining company, it noticed that the topographic map showed a cliff separating the two deposits, which significantly increased the complexity of the conveyor design. The geologists familiar with the site pointed out that there was no such cliff. The reason for the error was that the two maps supplied to the contractor were developed using early surveys, based on two different sets of reference points whose location and elevation were only approximate. These topographic maps were adequate for early exploration and as long as the two deposits were considered separately. A detailed aerial survey of the entire area should have been completed before initiating any detailed engineering.

This example illustrates the criticality of reliable information throughout the entire life of a project and the importance of physically inspecting the sites.

project manager takes responsibility to obtain reliable topographic maps or digital terrain models. These maps must cover not only the areas immediately surrounding the future mine site but also those needed to locate, design, and cost facilities that must be built outside the mine site for access roads, electricity supply, dams, water supply, and so forth.

The modeling of topography has been greatly facilitated by the introduction of aerial surveys, global positioning systems, and geographic information systems. However, modelers should remain aware of difficulties encountered when surveying steep topographic features or when in shadow areas where satellite access is obstructed. Dense vegetation can impede determination of earth surface elevation by aerial survey. Decisions must be made early concerning the coordinate system that will be used. Failing to do so can result in survey information collected in one part of the project being incompatible with that from another part.

GEOLOGIC MODEL: DOMAIN DEFINITION

Modeling a mineral deposit is a two-phase process. First the deposit must be divided into domains according to criteria designed to satisfy mining, processing, and environmental design requirements. The second phase consists of estimating the relevant properties of the deposit within each domain. Depending on the amount of information available, average values are assigned to each domain, or the domains are divided into blocks of fixed or variable size that are valued individually. Hence, the term *block model* is often used when referring to the resource model.

The criteria that are used to determine the need for domains are economically defined. The location and size of domains that satisfy these criteria are geologically defined. To design a mine, the mining engineer will need a deposit model that specifies the following:[1]

- Location and grade of economic and deleterious elements
- Geotechnical rock characterization as necessary to develop stable and safe mine plans
- Geometallurgical properties of mineralized material
- Specific gravity of material expected to be mined, including ore and waste
- Environmental characteristics of material to be mined and wasted or stockpiled

Domains will need to be defined for each one of these variables. In addition, the metallurgist will have specific requirements as needed to develop a process and to determine how a given process will perform during the life of the project. This involves information that the mining engineer also needs, including detailed information from which the geometallurgical properties of the mineralized material can be determined:[2]

- *Ore type*, a broad term generally understood to define domains within which the same metallurgical response is expected. As an

1 Additional geologic information that mining engineers may require is discussed in Chapter 14.
2 Additional geologic information that process engineers may require is discussed in Chapter 15.

example, porphyry copper deposits may be divided into leached, enriched, and primary domains or between oxide, transition, and sulfide domains.

- *Mineralogy*, such as percentages of pyrite, chalcopyrite, and bornite in a copper deposit
- *Geochemistry*, to determine the presence and quantity of potential co-products and deleterious elements

Geometallurgical characterization must not be restricted to the higher-grade parts of the deposit. It must include low-grade material, gangue, and waste that will be mined with the ore and sent to process. Initially, which material will be processed is only roughly defined. Cutoff grades are only broadly estimated during scoping studies and will change as feasibility studies are completed as well as during the life of the project.

Geologic information concerning waste and low-grade material is needed to design environmentally sound facilities, including waste dumps, stockpiles, tailings dams, and other earth structures. This information forms the foundation of the environmental model, which is discussed later.

Structures, mineralogy, alteration, lithology, and grade are typical geologic characteristics that are used to define domains. This characterization may or may not be sufficient. Early communication between geologist, mining engineer, and metallurgist are needed to ensure that domains are defined appropriately.

The number of domains depends on the complexity of the deposit. The accuracy with which the domains are defined increases when more data become available. A domain is meaningful only if there is enough information to define its location and internal properties. While in some cases the same domains might be used to model a number of variables, this should not be assumed. In a polymetallic deposit, such as one that contains copper, gold, and zinc, the controls of mineralization may be significantly different for each metal.

RESOURCE MODEL: BLOCK MODEL

The resource model is primarily understood to be one that represents in three dimensions the expected location, tonnage, and grade of material that could potentially be mined and processed. As such, this model represents the most important factor in the evaluation of mineral deposits. To create this model, the deposit is divided into blocks and each block is estimated separately. Like all other models, the resource model, also known as the grade model or block model, is not a perfect representation of the state of nature. Estimation errors cannot be avoided, which are directly translated into uncertainty concerning the profitability of the project.

Geostatistical Modeling

The information needed to develop the resource model consists of the geological domains and the grades obtained from samples whose size and location are known with reasonable certainty. The grades that are to be estimated are those required by mining and metallurgical engineers to make reliable decisions. The block grades are estimated by interpolation between drill holes, and the error of estimation increases when the distance between block and samples increases.

Geostatistics are typically used to estimate the grade distribution within appropriately defined geologic domains. These domains must be such that assumptions of stationarity, or constant statistical properties, can be reasonably made. The geostatistical methods are broadly defined as "kriging" (Journel and Huijbregts 1978).[3] There is, however, a very large and continuously increasing number of so-called kriging methods, and one must not assume that because kriging was used, the resulting model is sound.

The grade estimation methods are based on the premise that if we mix two samples, defined by their tonnage t_1 and t_2 and their grade x_1 and x_2, respectively, the average grade x_0 of the total tonnage $t_0 = t_1 + t_2$ is the weighted average:

$$x_0 = (t_1\, x_1 + t_2\, x_2) \,/\, (t_1 + t_2)$$

3 The term *kriging* was stamped by Professor Georges Matheron to honor Dr. D.G. (Danie) Krige, who first developed geostatistical methods to estimate the gold distribution in the Witwatersrand mines of South Africa.

The expected grade at a specific location is estimated as a weighted average of the grade of surrounding samples. If only two samples are considered, the following formula is used:

$$\text{expected grade: } x_0 = p_1 x_1 + p_2 x_2$$

$$p_1 + p_2 = 1$$

The weights p_1 and p_2 are estimated from the location of the samples with respect to the block being estimated and the geostatistical properties of the sample grades. The most significant geostatistical properties are the frequency distribution and variogram[4] of the known sample values. The variogram is a statistic that quantifies how the difference between sample values varies when the distance between samples increases. It measures the continuity in mineralization and the extent to which this continuity differs depending on the directions being considered.

Let u represent the location of a sample of grade x(u) and u + h the location of a sample of grade x(u + h), where h is the vector distance between the two samples. The variogram is calculated using the following formula:

$$\gamma(h) = \Sigma[\, x(u) - x\,(u + h)]^2 \,/\, [2\, n]$$

The sum applies to all pairs of samples that are separated by the same vector distance h, and n is the number of such pairs.

The geostatistical methods that apply to estimation of co-products (e.g., molybdenum in a copper-molybdenum mine) or contaminants (e.g., arsenic) belong to the same family of linear methods that apply to the grade of the main component (e.g., copper). However, the controls of mineralization, and therefore the geologic domains and geostatistical parameters that apply to co-products or contaminants, are not necessarily the same as those that apply to the main product.

There is a plethora of geostatistical methods with each one being best suited to answer specific questions. All methods have strengths and weaknesses. Ordinary kriging (OK) is often used to develop 3-D grade models. Under normal circumstances, if many surrounding samples are used to estimate a

4 The proper terminology is "semivariogram." However, the abbreviation "variogram" is now commonly used.

block grade, the OK estimate will be conditionally unbiased: If we consider all the blocks that are estimated at, say, 1.0% Cu, their average actual grade will be 1.0% Cu. Some blocks will be higher grade, some lower, but the average will be 1.0% Cu. This property is very important when separating ore and waste during mine operations. If OK is used as part of the ore control procedure, the grade of material received by the processing plant should be equal to the grade estimated by the mine.

To be conditionally unbiased, OK must underestimate high-grade areas and overestimate low-grade ones. This is known as the smoothing effect. Smoothing becomes increasingly significant when the distance between samples increases. For this reason, OK is not an appropriate method to develop a deposit model that will be used to develop mine plans and estimate tonnage and grade above cutoff grade. The information used to separate ore from waste is different from that available when the resource model is being developed. Geostatistical methods other than OK must be used to model mineral resources for estimation of the economic value of a mineral deposit.

Influence of Mine Selectivity on Resource Models

The information that is available to decide whether a deposit can be mined economically is usually obtained from widely spaced drill holes. From this information, it might be possible to estimate with reasonable confidence the average grade of large panels or groups of blocks within each geologic domain. But the grade of the small blocks that will be mined selectively[5] remains elusive. When mining the deposit, some amount of selectivity will take place. Only that part of each panel that exceeds a given cutoff grade will be sent to be processed; the rest will be stockpiled or wasted. This selectivity is made possible by sampling closer-spaced drill holes or blastholes. Evaluation of the deposit's economic potential requires estimation of tonnages and grades that will be selectively mined above a variety of cutoff grades.

The most difficult question faced by deposit modelers is as follows: How can I model the effect of selective mining when I do not have the closely spaced information that will be used during ore control? An approach that is seeing

5 The size of the smallest blocks that can be mined selectively is known as the selective mining unit, or SMU.

increasing popularity consists of using a very small number (as few as three or four) of samples to estimate blocks from exploration drill holes. Using this method, it is theoretically possible to develop a resource model that reasonably estimates total tonnages and grades that will be selectively mined above a given cutoff grade within large areas, such as those corresponding to one-year or longer time periods. But such a model is defective in many ways. The results are conditionally biased: Blocks near high-grade samples are overestimated, and blocks near low-grade samples are underestimated. Mine plans will target the high-grade blocks and will be invariably disappointed when mining these blocks. Conversely, areas expected to be below cutoff grade will contain some low-grade ore.

There are geostatistical methods that can be used to quantify the effect of selective mining without creating conditional biases. These methods cannot predict which part of a block will be above cutoff grade, but they can predict which proportion of the block will be above cutoff grade and the grade of this proportion. Deposit simulation is one of these methods.

Use of Deposit Simulation

Common practice consists of developing a single geostatistical model and designing mine plans, production schedules, and cash flows on the basis of this model. Even when only one grade model is used, the time and effort necessary to develop a feasible production schedule and estimate the corresponding cash flow is considerable. Unfortunately, a geostatistical model is only one possible representation of what the properties of the deposit might be. It is extremely unlikely that a production schedule that was optimized using a single model will be optimal, or even feasible, when mining takes place.

Furthermore, many resource models are conditionally biased, with areas surrounding high-grade samples being overestimated while areas near low-grade samples are underestimated. If ignored, these biases strongly influence the mine plan and result in misconceptions concerning achievable production schedules and cash flows. Geologic domains can be used with various successes to limit the extent of conditional biases. Parameters that are used to define the location and extent of such domains can also create biases or unrealistic models.

For these reasons, stochastic simulation of resource models is becoming increasingly favored as a way to model the effect of selectivity and develop better mine plans. Methods to simulate numerous possible realizations of the same deposit are commonly available and increasingly used to assess risk associated with a single mine plan. New algorithms are being developed, which are aimed at finding the mine plan that will result in the best economic return when all possible realizations of the resource model are taken into consideration (COSMO n.d.). Programs to take into account multiple simulated realizations of the same deposit to develop the optimal mine plan are only beginning to be used commercially.

Competence of Modeler

In some companies, mining engineers develop the resource model with input from geologist and metallurgist. In others, geologists are responsible for the model, with input from mining engineers and metallurgists. In either case, the resource modeler must qualify as a competent person (Chapter 4), which requires understanding of geology, geostatistical theory, and the practical aspects of resource modeling.

Geologic input must play a primary role in defining the domains within which geological and statistical stationarity[6] can be reasonably assumed. The geologist and geostatistician must review results obtained, not only once the model has been completed but continuously when the geostatistical analysis progresses. It is important to verify that the statistical properties of the deposit, derived from geostatistical analysis, are in agreement with what is expected given the current understanding of the controls of mineralization. If mineralization is expected to be high-grade along a fault zone and to decrease when the distance from the fault increases, the variogram should reflect the same trends. Any disconnect between geology and geostatistics must be explained. Either one or both may be wrong. Possible explanations may include incorrect definition of geologic domains, errors in variogram calculation, or insufficient data to confirm either geologic or geostatistical model.

6 Statistical stationarity implies that within the specified domain, the statistical properties of the deposit—such as mean, variance, autocorrelation, and so on—are constant.

Geostatistical theory is complex, and this complexity is increasing as new modeling methods are being developed. At the same time, the computer systems that are made available to practitioners by software companies are increasingly easy to use and can give an answer without requesting much input from the user. This combination of theoretical complexity and ease of use has advantages and disadvantages. Users with limited or no theoretical training can use the software, but they might not have the wherewithal to determine whether the results obtained are reasonable or why they are not. Conversely, users who fully understand theoretical geostatistics but have limited or no understanding of constraints imposed by geology and mining will be able to use the most complex algorithms while developing unrealistic models.

GEOMETALLURGICAL MODEL

The geometallurgical model is a 3-D model of the properties of the deposit that will influence the processing plant performance. It is aimed at predicting plant throughput, recovery, power consumption, flotation concentration ratio, and quality and quantity of output. The model must help predict how these variables will change during the project life as different parts of the deposit are being mined. During early evaluation, the metallurgist, geologist, mining engineer, and deposit modeler must jointly determine which information will be needed, the most effective way to obtain this information, and how to develop a meaningful model. Information taken into account to develop a 3-D geometallurgical model may include rock type, mineralogy, alteration, grade distribution, geochemical data, and hardness. Metallurgical tests (flotation tests, leachability tests, grinding tests, etc.) must be completed within appropriately defined geologic domains. The test results are used to determine which geological properties are most likely to influence plant performance and to refine the definition of geometallurgical domains.

Modeling methods such as kriging, which are applicable to the development of 3-D grade models, are not necessarily applicable to geometallurgical modeling. The reason is that metallurgical properties, such as recoveries, are not additive. If t_1 tons of material with grade x_1 are added to t_2 tons with grade x_2, the average grade is the weighted average x_0 calculated as follows:

$$x_0 = (t_1\, x_1 + t_2\, x_2)/(t_1 + t_2)$$

However, if the expected metallurgical recoveries from the same materials are r_1 and r_2, respectively, the average recovery r_0 is not the tonnage-weighted average of the recoveries. The following formula may be applicable where recoveries are weighted by metal content:

$$r_0 = (t_1\, x_1\, r_1 + t_2\, x_2\, r_2)/(t_1\, x_1 + t_2\, x_2)$$

The unknown in situ grade x_0 at a specified location can be estimated as the weighted average of the known grade of surrounding samples:

$$\text{expected grade: } x_0 = p_1\, x_1 + p_2\, x_2$$

$$p_1 + p_2 = 1$$

The weights p_1 and p_2 are determined taking into account the continuity in grades, as measured by the variogram. One could estimate the expected recoverable grade q_0 at the same location using a formula of the same type:

$$\text{expected recoverable grade: } q_0 = p'_1\, q_1 + p'_2\, q_2$$

$$q_1 = x_1\, r_1$$

$$q_2 = x_2\, r_2$$

$$p'_1 + p'_2 = 1$$

One might assume that the recovery at the specified location can be estimated by the following ratio:

$$\text{expected recovery} = (\text{expected recoverable grade})/(\text{expected grade}) = q_0/x_0$$

Although this approach may appear logical, it should not be used, as it is statistically, geologically, and metallurgically not valid. Statistically, the expected value of a ratio is not equal to the ratio of the expected values. Geologically, the statistical properties of grades and those of recovery are not the same. For example, in a gold deposit in which mineralization is controlled by vertical structures, recovery might be controlled by horizontal intensity of oxidation. The weight p_i given to the grade x_i is likely to be different from the grade p'_i given to the recoverable grade $x_i\, r_i$, and as a result, meaningless values of the ratio q_0/x_0 might be obtained.[7] Metallurgically, mixing material that has an

7 Recoveries calculated as the ratio q_0/x_0 can be higher or lower than any of the recoveries in the areas of interest. Recoveries greater than 1.0 can be obtained.

expected low recovery with material that has a high recovery could increase or decrease the overall recovery in a way that the preceding formulas do not recognize.

Many challenges abound in developing 3-D models of metallurgical recoveries. There are similar challenges with most metallurgical properties of mineral deposits, including mill throughput, power consumption, or concentration ratio. Development of appropriate geometallurgical models requires the following:

- The modeler must be a competent person with appropriate experience in geometallurgical modeling as it applies to deposits of the type being considered.
- The geologist, metallurgist, and modeler must review the nature of the models that are required, the information that is needed to develop these models, and the formulas that might be used to estimate the 3-D distribution of the geometallurgical properties of the deposit.
- Metallurgical tests tend to be expensive. In some deposits, whole-rock analyses of major and minor elements can be obtained at reasonable cost, and the results can be used to predict mineralogy and metallurgical response.
- In some cases, estimation of average geometallurgical properties within domains or within large panels may be sufficient. In more complex cases, a block-by-block estimate of geometallurgical properties might be necessary.
- Once completed, the geometallurgical model must be validated, not only by the geologists but by the metallurgist to make sure that the results are realistic, metallurgically speaking.
- The limitations of the models must be well understood and shared with those who will use the models. How much testing is completed should be a function of geologic complexity, costs, and risk taken as a result of limited ability to predict plant performance.

ENVIRONMENTAL MODEL

Whenever material is mined, moved, stockpiled, or otherwise exposed to the elements, there is a risk that these activities will have environmental consequences. Of particular interest are the geochemical properties of this material and its acid-generating and acid-neutralizing potential. The environmental model is aimed at determining the potential for water contamination and acid mine drainage resulting from leaching of heavy metals and other deleterious elements. The model is required for waste management, dump design, environmental control, and monitoring.

A waste characterization model is needed to give mining and environmental engineers information about the material that will be mined and placed in waste dumps or low-grade stockpiles, as well as material to be used for construction of tailings dams, earth berms, and other earth structures. Environmental characterization is also needed for soil, subsoil, and rock that will be exposed or removed during mine pre-stripping and construction of surface infrastructure, including processing plant, tailings storage facilities, leach pads, waste dumps, offices, access roads, and so forth.

Environmental engineers define information that is needed to make sound decisions, including sample types and testing methods. With the geologist, they specify where samples should be taken. The competent person acting as project environmental engineer must be aware of, and take into consideration, all regulatory requirements concerning rock characterization, design of facilities, and water quality monitoring as required to obtain the necessary construction and operating permits. In some cases, defining average environmental characteristics may be sufficient; in others, extreme values are more important. Sampling and modeling methods must vary accordingly.

The cost of environmental characterization can be controlled by limiting data collection and analysis to the minimum required by regulations. All those whose decisions depend on availability of reliable environmental information, especially mining, processing, and infrastructure engineers, should assess the risk taken by following this cost-limiting approach. For example, insufficient waste characterization can result in locating potentially acid-generating waste dumps in locations where leakage could have catastrophic environmental

GEOTECHNICAL CHARACTERIZATION AND INFRASTRUCTURE DESIGN

Case 1

In one underground mine, the production shaft was located to give easy access to high-grade mineralization and minimize initial capital and operating costs. Geotechnical conditions had been tested at the shaft location but were found worse than expected when the shaft reached the mineralized zone. Geotechnical information obtained during shaft sinking indicated that mining high-grade material close to the shaft was not feasible. It would result in long-term shaft instability and eventually prevent access to more distant mineralization. Once targeted production was reached, both capital costs and initial operating costs had significantly exceeded budget estimates. Detailed modeling of geotechnical conditions should have been completed before finalizing mine plans and production schedules, not only where the shaft was to be located but also in nearby areas. For geological reasons, mineralized zones are often less geotechnically stable than waste zones.

Case 2

In this example, the location of the primary crusher was chosen to minimize the cost of transporting material from mine to crusher and from crusher to processing plant. When bedrock was exposed to build the crusher foundation, it was realized that the chosen location was partly on top of a major fault, which was unsuitable for construction. The cost of relocating the facility could have been avoided if appropriate geological modeling of the area had been completed.

Case 3

In this case, engineers were designing a leach pad to treat ore from a low-grade-oxide gold mine. The only suitable area was in a relatively flat alluvial valley between steep hills. Preliminary geological mapping showed a number of water-saturated bogs in the area. Determining the location, size, and properties of all bogs and replacing the boggy material with rock was considered too time-consuming and costly. A geotechnical study showed that loading the pad on top of the bogs would create instabilities unless a specific loading sequence was followed, which would require careful supervision. Unfortunately, assuming continuous 24-hour engineering supervision was not realistic. Three months after leaching started, a major pad failure occurred. The pad had to be off-loaded, all boggy material had to be removed, and the pad had to be rebuilt before gold production could resume. The likelihood of pad failure and the cost of remediation if it failed had not been given sufficient consideration. The risk taken when accepting the original operating assumptions had been significantly underestimated.

consequences. Minimizing current costs can result in inflated future costs, including costs of reclamation and closure.

GEOTECHNICAL MODEL

Geotechnical information is needed for a variety of purposes. In open-pit mines, mining engineers need geotechnical information to design stable pit slopes and bench heights, and to locate haulage ramps. In underground operations, geotechnical information is needed to determine which mining method is technically and safely feasible, to design stopes and other openings, and to determine support systems that will be required. Geotechnical information is needed before finalizing the location of shafts and underground accesses. The location of primary crusher, processing plant, tailings storage facilities, leach pads, waste dumps, and other surface facilities must take into account underlying soil, subsoil, and bedrock geotechnical properties. Geotechnical information is needed to determine blasting requirements, and for process engineers to design comminution circuits and estimate throughput.

The type of geotechnical information that is needed may vary depending on whether information is to be used for open-pit or underground mine design; to estimate blasting requirements, including blasthole spacing and powder loading; to design comminution circuits; or to locate surface facilities. Mining engineers need reliable information concerning the location, orientation, and physical characteristics of fault zones. Typical geotechnical information that the geologist must document and model include rock quality designation (RQD) and rock mass rating. Process engineers may need additional rock characterization such as point load index and Bond work index.

Compatibility between geological and geotechnical models must be ensured. Effective communication is essential between those who will gather geotechnical information and those who will use it. Geotechnical logging, testing, and modeling requires special expertise, which is the prerogative of a geotechnical engineer working with the project geologist. In some cases, average geotechnical properties within specified domains may be sufficient; in others, values above a specified level may be more relevant. For mine design, the location and extent of fault zones with low RQD is highly relevant. Small amounts of mineralized material with a much higher than average Bond work index can significantly impact mill performance.

When designing open-pit mines, the geotechnical risk can be quantified by estimating factors of safety and probability of slope failure, and by calculating the cost of failure. The geotechnical risk is expected to vary over time as the pit deepens. Changes in probability of failure can be estimated by analyzing pit slopes at different times during the planned mine life. To avoid underestimation, consideration should be given to the probability that previous mining activities created new zones of weakness in the pit walls that did not exist before operations started.

In underground mines, what is mined at a given time puts constraints on what can be mined in following years. Although all mining operations leave lower-grade material behind, sustainable management of resources demands that current designs take into account potential future needs. This is particularly important when a deliberate decision is made to mine only the highest portion of potential stopes. Such activities might be economically justified, whether for net present value (NPV) optimization[8] or to satisfy short-term cash requirements. However, a result is likely to be sterilization of lower-grade material whose safe access will no longer be feasible at a later stage, thus reducing life-of-mine and related (although uncertain) future opportunities.

HYDROLOGICAL MODEL

Hydrological models must reflect surface and subsurface water quantities, qualities, and flows. They must take into account current usage and be suitable to predict the impact of future usages. Baseline studies must take into account the 3-D physical space, including interaction between groundwater, surface water, and the atmosphere. A characteristic of water is that it flows: Information that must be modeled must include vector information, the direction in which water is flowing, and the quantity and quality of flow. Conditions change over time, from month to month, during dry and wet seasons, and as a result of

8 Optimal cutoff grades and mine plans, defined as those that optimize the net present value, depend on the discount rate. High discount rates are used to reach such objectives as minimizing political risk. In extreme cases, sensitivity of mine plans to discount rate should be investigated before making a final investment decision.

INCOMPLETE SURFACE HYDROLOGICAL MODEL

The surface hydrological model must account for rivers, streams, lakes, swamps, wells, irrigation ditches, and other easily observed surface features. Nevertheless, it is easy to miss features that may seem insignificant to the eye of a hydrologist or a mining engineer not familiar with the local culture. For example, when mapping waters in the high Andes, irrigation channels whose width might be as small as 30 cm and in which water flow may appear insignificant, should not be ignored. Such channels, which may date from the Inca period, bring water to downstream habitations and are used to irrigate vegetable gardens and supply drinking water to animals and possibly humans. These channels are easily ignored when building roads or crossing undeveloped land, with dire consequences when farmers initiate violent protests as a result of having lost access to what was a reliable and vital water supply.

longer-term trends or extreme weather events. Taking changes in both space and time, a hydrological model is four-dimensional (4-D).

The model must be usable to predict the need for mine dewatering, the impact that such dewatering will have on current surface and underground waters, and how current users will be affected. Disposal of mine water, whether via surface streams or underground aquifers, will influence water quantity and quality in a way that needs to be predicted. The processing plant will need water as input. Process water that cannot be recycled will be disposed of in tailings storage facilities, which need to be monitored for leakage. The hydrological model is needed to determine possible sources of appropriate process and potable waters and to assess the impact of consuming these waters.

A detailed survey of surface waters, including springs, streams, rivers, lakes, swamps, dams, irrigation channels, and flood zones, must be completed before designing and locating surface facilities. Mine closure must take into account the hydrological conditions expected to prevail at the end of the mine life, the need for remediation, and the medium- to long-term impact that pit lakes and flooding of underground openings will have on quality and quantity of future water conditions.

Geological and hydrological information is needed to develop a hydrological model. Aquifers are modeled using data obtained from wells and drill holes.

Data gathering should start during early exploration, documenting the elevation at which water is encountered when drilling for geological information. As the project progresses, dedicated holes are located to complete pumping tests. Observations made in pumped holes and surrounding observation holes form the basis on which the 4-D hydrological model is developed.

The impact of deep mining on aquifers can cover large areas. The location of the test holes takes into account the regional geology and rock properties, including rock porosity and permeability, presence of aquitards, water saturation in fault zones, and so on. Detailed geologic knowledge is generally limited in these typically nonmineralized areas. The cost of aquifer testing and the size of the areas to be tested contribute to limiting the number of data points available to model current water conditions and predict future ones.

The small number of hydrology holes and the complexities of underground hydrology can result in significant uncertainty concerning mine dewatering needs, impact of mining on shallow and deep aquifers, properties of water pumped from the mine, and the impact of mine flooding after closure. Considering that the impact of water on the environmental and social bottom lines can be very significant, continuous monitoring and impact remediation are needed during the entire project life and after closure.

SOIL, SUBSOIL, AND BEDROCK MODEL

A soil, subsoil, and bedrock model is required as part of a baseline study to determine biological properties, to ascertain the need for pre-stripping and stockpiling for future environmental remediation, and to assess the risk of soil erosion and contamination. This information is also needed as an input to civil engineering activities such as infrastructure location and design. This model must cover areas where mining activities will take place as well as all areas where infrastructure is likely to be built.

RISK MANAGEMENT AND MODEL BUILDING

Both theory and practice of model building vary considerably, depending on what is being modeled and how the model will be used. Multiple books and technical articles have been published, and many specialized conferences and short courses are organized to disseminate knowledge. Universities and consulting companies conduct research to develop new theoretical and practical modeling methods. Consulting services and computer systems are available to in-house modelers. It is not the intent of this book to delve into these highly specialized fields. However, concluding remarks are useful concerning common sources of uncertainty, how this uncertainty can be quantified, and how the project risk can be managed.

Competence of Modeler

The person responsible for developing the necessary models must have appropriate expertise and must qualify as a competent person. Because modeling is highly specialized, different competent persons are likely to be needed to model geology, grades, geometallurgy, geotechnical characteristics, or hydrology.

Model Purpose

Before developing a model, it is necessary to understand how the model will be used. The grade model is likely to vary depending on the mining method. The geotechnical model required for underground mine planning is not the same as that for an open pit. Understanding of environmental rules and regulations is needed when developing an environmental model. The models that metallurgical engineers require are a function of the process that is likely to be used. No model should be developed before prior consultation with those that will use it.

Model Input

The information used to build models can be objective or subjective. Data are obtained by completing tests and measurements that are subject to errors. Subjective information includes that obtained from drill-hole logging. The quality of subjective information varies with the individual generating this information. Two geologists who are logging the same core may come to different conclusions. The samples that are used to estimate grades or metallurgical characteristics must be representative of those parts of the deposit that will

MISLEADING RECONCILIATIONS

In operating mines, reconciliation of earlier models with current operational results is one of the best methods to validate and continuously improve the resource model. One should take reasonable precautions to ensure that such reconciliations are meaningful. This requires a complete understanding of the source of information that was used to develop the original model and that with which it was compared. Reconciling an ore control model developed from blasthole grades with a recent deposit model that used both drill-hole and blasthole grades is meaningless. Unless gross errors were made in developing the models, the resource model (which includes blastholes) should show perfect reconciliation with the blasthole model.

be mined and processed. Models are only as good as the information used to generate them. Quality assurance and quality control (QA/QC) procedures should be in place and followed to ensure that data errors are controlled and quantified. This applies to both data entry and maintenance of databases.

Modeling Method

Modeling methods vary depending on the variable being modeled and how the model will be used. Methods must be judiciously chosen. Different methods apply to modeling categorical variables (e.g., boundaries between oxide, transition, and primary geological zones) and continuous attributes (e.g., grade and quality of mineralization). While the theory underlying some modeling methods might be relatively simple, that underlying more complex and presumably more accurate methods might be understood by only a small number of those who develop models. Theories are based on weak and strong assumptions. Being unaware of these assumptions means that the need to determine whether they are reasonably satisfied is ignored. This may result in inappropriate modeling methods being used and unreliable results being obtained. These remarks are applicable to any modeling method, whichever variable is being modeled. It is often advisable to initially develop several models of the same variable, using different methods, and to validate the results before deciding which method should be retained. The most appropriate method is not necessarily that which gives the most desirable results.

COMPATIBILITY BETWEEN MODELS

All models—including those of mineral resources, geotechnical properties, geometallurgy, hydrology, soil, subsoil, and rock characteristics—have one common denominator: geology. To minimize costs and avoid inconsistencies between models, information gathered to improve one model should be combined with information acquired for the other models:

- The geostatistical model, including the properties of the variogram used to develop the model, should be compatible with the geologist's understanding of the controls of mineralization.
- Information obtained when drilling geotechnical holes in waste zones surrounding a mineral deposit should be compared with that obtained by resource geologists within the mineralized zones. This may seem obvious, but resource geologists and geotechnical engineers are often located in separate buildings, sometimes kilometers apart. Situations do occur where the same fault modeled by two different teams are given different names and different probable locations.
- In sulfide deposits, the geometallurgical model must give a reasonably accurate representation of the extent of oxide, transition, and sulfide zones. The surfaces separating one zone from the other can be very complex. Poor definition can result in significant errors in prediction of plant performance and early cash flows. This risk is often underestimated, resulting in insufficient geologic information being obtained in those parts of the deposit scheduled for early production. Drilling short infill holes to characterize the first two years of production is an efficient way to significantly reduce start-up risk.
- Information considered necessary to develop a reliable hydrological model is a function of the extent to which the geology of the areas in and surrounding the mine is understood. This requires sharing of information between resource geologists who are logging holes for resource assessment and geotechnical characterization, and hydrologists who are interpreting pumping tests for hydrological modeling.

Simulation

While modeling is generally understood as the development of a single expected value for the variable being estimated, simulation is used to create a number of equally likely realizations of what the deposit properties might be. Each realization respects the data used to develop the model, while recognizing that there is uncertainty concerning values between data points. Simulations are particularly useful to determine uncertainty and risk. Mine plans and production schedules are often developed using a single model. One can measure the risk associated with this production schedule by testing the results using a number of possible realizations.

Model Utilization

Models are developed to satisfy requirements specified by those who will use them. Once a model is completed, a formal handover process should be followed, during which the model is transferred to its downstream users and these users are made aware of its strengths and weaknesses. Models are designed to answer questions. If questions change, models may have to be changed. Grade models and geotechnical information needed to design an underground mine differ from those needed to design an open pit. The grade model should take into account the selectivity to be achieved by the mining method. When analyzing a potential open-pit mine, the original assumption might be that 7-meter benches will be used. It may later be decided that 10-meter benches are more appropriate because this adjustment would result in lower costs and higher productivity. Evaluating this change is likely to require a new deposit model that reflects the lower selectivity.

Model Reconciliation

Whenever possible, models that were developed at a given point in time should be reconciled with observations made at a later stage. In operating mines, predictions made from drill-hole-based resource models should be compared with actual production realized from mined-out areas. The results obtained and lessons learned are used to improve the resource model. A complete reconciliation process should compare resource model, ore control estimates, material movement from mine to destination (mill, stockpile, waste dumps), quantity and quality of mill feed, product sold, and rejects.

Reconciliation should not be limited to the mineral resource model. The hydrological model is used to predict changes that will result from mine operations. The geometallurgical model is used to predict plant performance.

These predictions should be compared with actual results, such as water inflow in the pit or plant throughput and recovery.

Auditing

The economics of a project are first and foremost dependent on the mineral resource available. Models should be independently audited before major investment decisions are made. When choosing auditors, one must keep in mind that they are not of equal value. Some consulting companies are known to endorse their client's opinion whatever it might be, while others are known to be highly critical—sometimes overly so—and for making strong recommendations for changes that may or may not please their client. Public release of mineral resources and mineral reserves are subject to regulations that specify situations when independent audits are required. Stock exchange regulators are generally aware of which consultants can be trusted and which cannot.

CHAPTER 14

MINING ENGINEERING

I now come to the mode of working, which is varied and complex, for in some places they dig crumbling ore, in others hard ore, in others a harder ore, and in others the hardest kind of ore. In the same way, in some places the hangingwall rock is soft and fragile, in others harder, and in still others of the hardest sort.

Georgius Agricola, *De Re Metallica*, 1556

The role of the mining engineer is to define the most efficient way to extract material of economic value from the ground and deliver it to the processing plant. Mining efficiency must take into account the constraints imposed by geology, the requirements of the metallurgical process, capital and operating costs, and safety. Development of a mine plan and the production schedule must satisfy social, environmental, legal, and safety regulations as well as company objectives and values. The three bottom lines—financial, social, and environmental—must be assessed to compare various scenarios. A mine plan cannot be evaluated in isolation.

RELATIONSHIP BETWEEN MINING AND OTHER TECHNICAL DEPARTMENTS

The resource model is the foundation on which mine plans and production schedules are developed, and mine capital and operating costs are estimated. Tonnage and quality/grade of material delivered to the processing plant form the basis for estimation of revenues from sales. If the simple assumption of a linear decision-making process is made, the relationship between resource

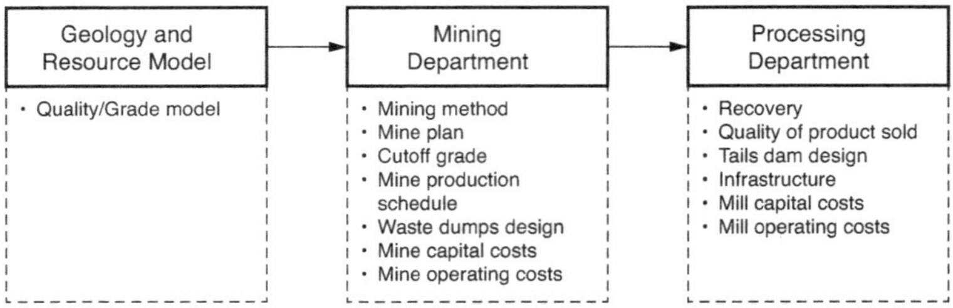

FIGURE 14.1 Linear (oversimplified) decision-making process in mining engineering

modeling, mining, and processing can be represented by Figure 14.1. The underlying assumption is that a three-dimensional (3-D) model of the quality/grade of material available to be mined is sufficient to develop a mine plan. Even if such a simple assumption is made, a meaningful production schedule cannot be developed without making additional assumptions concerning the geotechnical environment, what metallurgical process could be used, processing costs, metallurgical recovery, and value of product that will be sold.

Determination of the economic potential of a mineral project is an iterative process. Mining engineers must make decisions concerning mine plan, sequencing, cutoff grades, production schedule, available mill feed, stockpiling, and waste disposal that require detailed information about the location, tonnage, grade, and metallurgical properties of mineralized material over a number of possible cutoff grades. Mining decisions influence processing decisions, which in turn influence how the mine should be operated. Net present value (NPV) optimization requires analysis of different cutoff-grade strategies and production rates and determination of the corresponding capital and operating costs, which in turn influence the cutoff-grade strategy.

A more realistic representation of the mining decision-making process is shown in Figure 14.2. The figure illustrates the wide range of information that the mining engineer needs to develop mine plans and production schedules, and to estimate capital and operating costs.

The geology and resource models needed must include not only a 3-D representation of the grade/quality of material of potential economic interest but also geological, geotechnical, geochemical, geometallurgical, and hydrological information. Conversely, mining and processing methods influence

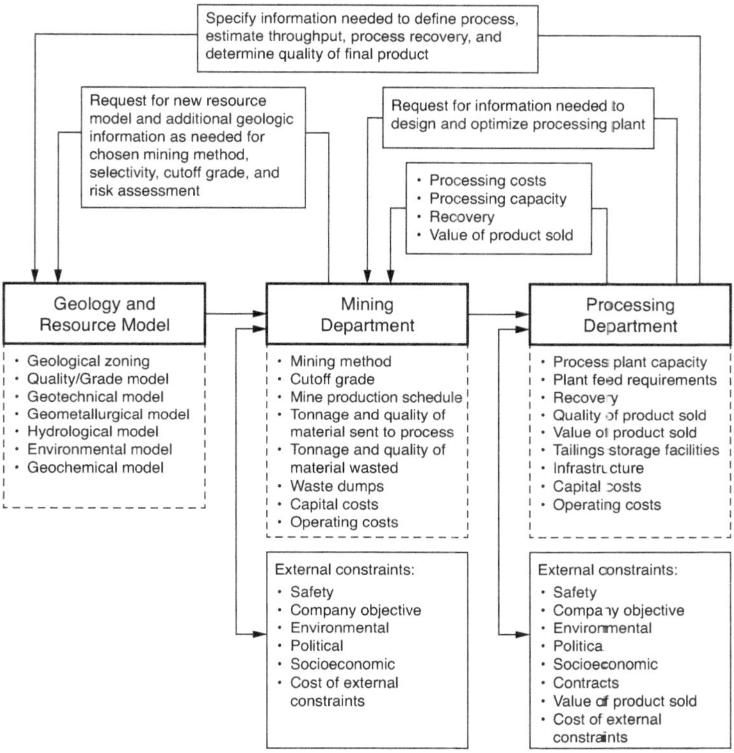

FIGURE 14.2 Complex (realistic) decision-making in mining engineering

the type of models that mining engineers require. Requirements are not the same for open-pit and underground mining and vary with mine selectivity. Geometallurgical and geochemical information that is needed differs when evaluating oxide or sulfide mineralization, and waste characterization has specific stipulations.

Geology is not sufficient to design mines and determine practical production schedules. Economic parameters are needed, including mining and processing costs, plant recovery, and value of product sold. A cutoff grade is needed to estimate quantity and quality of material that can be sent to the processing plant. However, cutoff grades and costs cannot be determined with confidence until a production schedule is suggested, which in turn defines mine and mill capacities, capital costs, and operating costs.

For these reasons, the decision-making process must be iterative. Rough assumptions and educated guesses are made during early evaluation. The need for additional or better information is determined. When information improves, new mine designs and production schedules are developed. Cash flows and project risks are reestimated. This process is repeated as project evaluation progresses from scoping study to pre-feasibility study and feasibility study. This chapter investigates the role of the mining engineering department, how mining decisions influence the utility of the project as defined by the three bottom lines, and which information is needed to make sound mining decisions.

ROLE OF THE MINING ENGINEERING DEPARTMENT

Mining engineers must make a wide variety of decisions and recommendations to ensure successful evaluation of a mineral resource. Only some of the items listed below are likely to matter in early-stage evaluation, before and during scoping studies; most of them will apply when reaching feasibility studies. The list is not the same for all projects and varies with project location, geology, type of mineralization, level of development, and so forth. It is the responsibility of the mining engineer to determine which item is of significance at which stage. However, all items listed here should be given due consideration before deciding that they do not apply. Tasks to be completed by the project mining engineer include the following:

- Review and be knowledgeable of all applicable rules and regulations as well as company objectives and values, including those related to safety, environmental compliance, and social responsibility.
- Review project information with the geologist, metallurgist, geotechnical engineer, and hydrologist.
- Ensure competence. The mining methods that are considered might be open pit or underground, a bulk method such as block caving, or a selective method such as drift and fill. A different level of expertise will be required from the mining engineer, depending on the method retained.

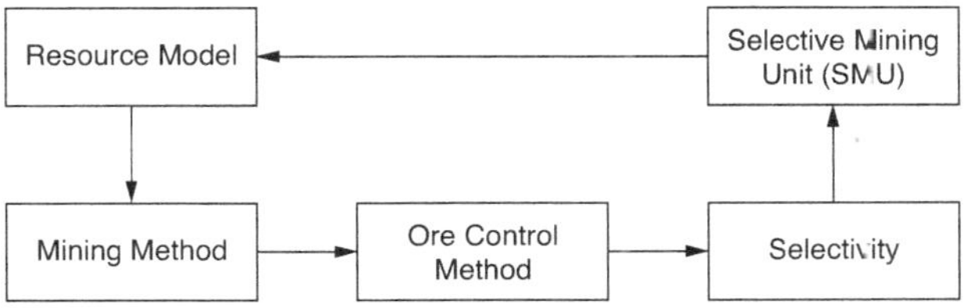

FIGURE 14.3 Relationship between resource model and mining method

- During the early stage of a project, up to and including the pre-feasibility study, more than one mining option or production schedule should be considered. A successful pre-feasibility study will result in a single option being retained, which will be the subject of a feasibility study.
- Determine which mining methods are applicable. Mining methods are chosen taking into account available geological, geotechnical, geometallurgical, and hydrological information. Safety considerations are a key concern.
- Define an ore control methodology. Ore control has a direct effect on mine selectivity and the definition of selective mining units (SMUs). Deposit models must take into account mine selectivity and the size of the SMU. Figure 14.3 shows how a resource model that was appropriate at a given stage may have to be modified when additional information is obtained concerning mining methods.
- Ensure the suitability of the resource models. The deposit models that are appropriate to study the feasibility of open-pit, underground, selective, and bulk mining methods are not likely to be the same, even if they are based on the same drill-hole information. Geotechnical information needed to design pit walls is not the same as that needed to design underground openings or comminution circuits. All models should be compatible because they all aim at representing, albeit in different ways, the fixed but imperfectly known properties of the deposit.
- Tonnage and grade of material that will be delivered to the processing plant must take into account ore losses and external dilution. Losses and dilution depend on geology as well as mining and ore control methods. In some instances, such as in open-pit mines with

MINE DESIGN AND GEOTECHNICAL RISK

Case 1

A company's best-practice guidelines recommended that for deep open-pit mines, two access ramps should be designed, thus minimizing the risk of loss of production resulting from ramp failure. A new mine was being designed and the pit depth was expected to exceed 500 m. On the basis of current metal prices, the project economics were marginal. However, the general consensus was that prices were exceptionally depressed and that by the time the mine would be operational, economic conditions would improve. To minimize stripping costs and improve economics, consideration was given to designing a pit with only one access ramp. To decide whether this was acceptable, two scenarios were compared. It was first assumed that the single ramp would not suffer any major failure. The cost savings resulting from one ramp instead of two were quantified. Then the assumption was made that the ramp would fail at some point during the life of the mine. The cost of such a failure was quantified, depending on when it would occur and how long it would take to remediate the situation. This cost took into account the cost of regaining access to the pit and loss of revenues resulting from interruption to production. Probabilities of failure were estimated, taking into account available geotechnical information, ramp location, pit depth, and possible weather events. After considering these scenarios, management decided that the risk associated with a one-ramp solution was not acceptable because of uncertainty concerning the geotechnical properties of the pit walls where the ramp would be located. The decision was made to postpone the decision for six months when additional geotechnical information would be obtained, a new ramp design would be completed, and the probability of failure would be reassessed.

Case 2

When designing open-pit mines, insufficient or incorrect geotechnical information can result in slope angles being flatter than necessary, resulting in substantial increase in strip ratio and mining capital and operating costs. Alternatively, the result could be the design of overly steep pit slopes, resulting in major slope failures, which could have been avoided if appropriate geotechnical information had been collected on time and properly analyzed.

Case 3

When designing underground mines, lack of appropriate geotechnical information can result in underestimation of safety conditions, overestimation of achievable productivity, and underestimation of mining costs. Mine development might be slower than planned, the size of the stopes might have to be reduced, and additional support might be needed to ensure safety. In extreme cases, mines have been shut down shortly after production started because of unsafe operating conditions. Alternatively, the result of poor geotechnical information can result in overestimation of costs and underestimation of achievable productivity, resulting in abandoning a project that could have been profitable.

disseminated mineralization, losses and dilution can be included in the resource model. In other situations, especially when significant structural controls are present, the mining engineer must estimate losses and dilution and include them in conversion of mineral resources to mineral reserves.

- Obtain information from the process engineer about plant capacity, constraints related to the quality of mill feed, processing cost estimates, metallurgical recovery, and value of product to be sold. This information is preliminary at the beginning of the project analysis and improves over time as definition of the mine production schedule improves. While the mine production schedules and quantity and quality of material delivered to the processing plant must take into account constraints imposed by the plant, the plant design is in turn a function of the quantity and quality of material that can be delivered by the mine.
- Develop mining cost estimates. Throughout the decision-making process, mine capital and operating costs must be estimated. Preliminary estimates are made at the beginning of the process, which become increasingly precise as the economic study progresses and more information is available. The precision and accuracy of the cost estimates are a key component of the project risk.
- Estimate the marginal cutoff grade on the basis of expected mining and processing costs, mine dilution, process recovery, and expected value of recovered product. The cutoff grade must take into account constraints imposed by process, environment, and market on the quality of material that can be processed economically. Initial cutoff grade calculations are based on order-of-magnitude estimates of input parameters. As the study progresses, these estimates are revised upward or downward. Cutoff grade estimation must take into account mining and processing method, losses, recovery, costs, and value of product to be sold. A detailed discussion of how to estimate cutoff grades is outside the scope of this book and can be found in *An Introduction to Cut-off Grade Estimation* (Rendu 2014).
- Develop a mine plan and production schedule. The production schedule shows expected quantity and quality of material to be mined on a yearly basis (or some other time frame), including waste

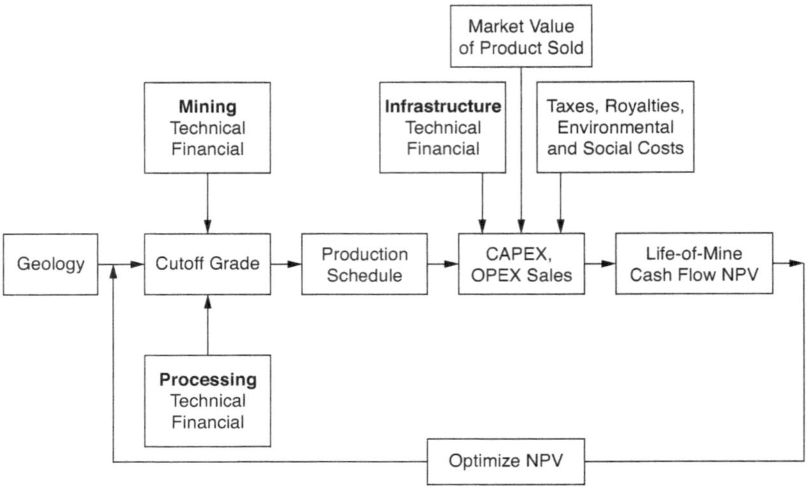

CAPEX = capital expenditure; NPV = net present value; OPEX = operating expense

FIGURE 14.4 Optimization of cutoff grade, production schedule, and cash flow

material, material directly sent to the processing plant, and material stockpiled for later processing. The mining rate is determined by deposit characteristics, mining method, geotechnical considerations, equipment selection, and safety considerations. Knowledge of, or realistic assumptions concerning the processing plant capacity and material type requirements, are needed before a mine production schedule can be developed.

- Maximize the project's NPV. The measure most commonly used to optimize mining decisions is the NPV of the cash flow that the project is expected to generate (Figure 14.4). The first production schedule is not likely to be optimal. If this schedule is based on a marginal cutoff grade, it is likely to result in overdesign of the mining and processing facilities and overestimation of capital costs. To maximize NPV, one must attempt to use a higher cutoff grade early in the mine life while stockpiling low-grade material. Except when technical constraints and costs are such that higher-grade material cannot or should not be prioritized, the marginal cutoff grade is only used at the end of the mine life. When calculating this cutoff grade,

MINE DESIGN AND WATER MANAGEMENT

A new porphyry copper open-pit mine was being designed, with a pit reaching 900 m over a 20-year period. The mine was located in a tropical area where considerable rainfall could be expected during half the year. A preliminary design was completed to mitigate flooding during the rainy season, including water diversion channels and pumping facilities. To better assess the risk associated with weather patterns, a detailed review of historical rainfalls was completed and adjusted for the likely consequences of global warming and climate change. The current design was found to be inadequate. Diversion channels would fail to prevent pit flooding at least one week a year, and catastrophic flooding, resulting in destruction of water control facilities and one-month production interruption, should be expected once every five years.

It was determined that modifying the water diversion and dewatering systems could reduce yearly production interruptions from one week to one day. Catastrophic pit flooding could not be fully avoided, but redesign of flood control facilities and an increase in dewatering capacity could reduce the duration of pit closure by half. The mining rate was also increased to allow stockpiling of lower-grade material during the dry season, which would be processed during the wet season.

all the costs that will be incurred by continued operation—including capital and operating costs, and fixed and variable costs—must be taken into account, as well as the benefit, if any, of postponing closure.

- Locate and design waste dumps and stockpiles, taking into account the need for environmental compliance. Such designs require that the resource model contain information concerning waste and stockpiled material, including geochemical characterization as well as acid-generating and acid-neutralizing potential.
- Develop a mine closure plan, including pit and underground reclamation, waste dumps, haulage roads, and other infrastructure. Continuous reclamation should be considered to minimize end-of-life reclamation costs.
- Assess water consumption and mine dewatering requirements. Locate and design sources of water and disposal of excess water.

- Estimate mine equipment requirements, including drills, loaders, haulage trucks, dozers, scrapers, water trucks, light vehicles, and so on. The size of mining equipment, when combined with the ore control method, are key determinants of the achievable mine selectivity.
- Estimate staffing requirements, including that needed for mine operations and maintenance, mine planning, ore control, training, safety, and overhead functions.
- Estimate mine capital costs, including
  - Capitalized mine development (pre-stripping, shaft sinking, initial underground envelopment, site preparation for waste dumps and stockpiles, haulage roads);
  - Mine equipment;
  - Maintenance shops and explosives storage;
  - Office space and equipment, including information systems; and
  - Capital expenditure expected to be needed during the life of the mine, including sustaining or expansion capital.
- Estimate mine operating costs, including
  - Staffing (management, mining engineering, mine operations, maintenance, safety, training, security);
  - Mine operating costs, including drilling, blasting, loading, trucking, dewatering, waste and stockpile management, consumables (fuel, electricity, water, explosives);
  - Equipment maintenance costs;
  - Ore control costs (including sampling, laboratory, in-pit and underground ore control, and surface requirements); and
  - Overhead costs, including mine site and head office.
- Estimate mine closure costs, such as
  - Continuous reclamation costs;

- One-time closure costs at the end of the mine life; and
- Cost to perpetuity such as may be incurred to treat water effluents from mine or waste dumps.

- Convert mineral resources to mineral reserves. As indicated in Chapter 4, a mineral reserve is the economically minable part of a measured and/or indicated mineral resource. It includes diluting materials and allowances for losses, which may occur when the material is mined or extracted. It is defined by studies at pre-feasibility or feasibility level, as appropriate, that include application of modifying factors. Such studies demonstrate that at the time of reporting, extraction could reasonably be justified. While a geologist is usually the competent person responsible for the estimation of mineral resources, the project mining engineer is the competent person responsible for estimation of mineral reserves. Since mineral reserves require completion of a pre-feasibility or feasibility study, the mining engineer must ensure that there is sufficient confidence in the results of this study before declaring a mineral reserve. This confidence must take into account not only mining input but all other technical and economic input that is included in the economic study.
- During a pre-feasibility study a range of mining, processing, and infrastructure options must be analyzed to the level where a preferred option can be chosen. Such options may consider various mining methods, bench heights, mining rates, plant designs, processing rates, and so forth. Each option must define technical and economic parameters. The reported mineral reserve should be that which corresponds to the preferred option.

This partial list illustrates how complex mining technical and financial decisions can be. Mine planning is, by necessity, an iterative process. Preliminary assumptions are made, a production schedule is developed, the original assumptions are refined, and a new schedule is developed with increasingly reliable estimates of both input and output variables. The process is repeated until sufficient confidence in the mine plan has been reached (Figure 14.5). Coordination between mine planning and process design is important. A

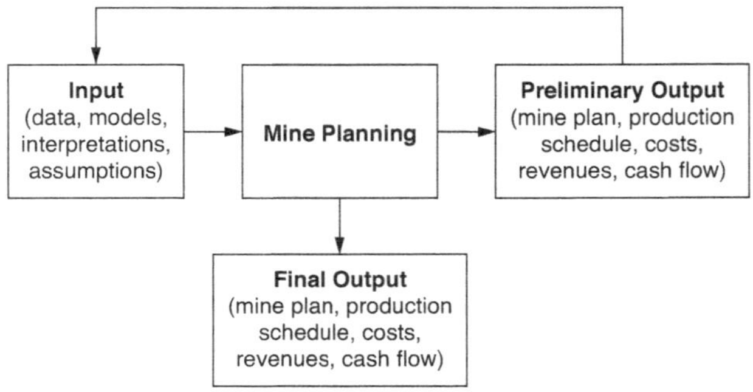

FIGURE 14.5 Role of the mining engineering department

final mine plan developed using detailed mine-related information can be meaningless if process information is of much lower quality.

MINING DECISIONS AND THE FINANCIAL BOTTOM LINE

Mining decisions influence the project utility through their impact on the financial, environmental, and social bottom lines. To properly value a project, the mining engineer must be fully cognizant of the information needed to evaluate the three bottom lines. This entails ongoing and effective communication between the mining engineering department and the other departments (geology, processing, environmental, legal, public relations, accounting, procurement, sales, taxes, etc.) whose requirements impose constraints on mining and whose activities are constrained by mining decisions.

The financial bottom line is measured in dollar terms, and its value is assessed by a number of criteria, including NPV, internal rate of return, yearly cash flow, capital requirements, payback period, unit operating costs, and other factors that may be weighed differently depending on the company's priorities and needs. These estimates are based on the cash flow that the project is expected to generate. The main contribution of the mining department to the financial bottom line is the yearly mine production schedule and the capital and operating costs that will be incurred to maintain this production.

The direct contribution of the mining department to cash flow is entirely negative. However, decreasing mining costs does not necessarily mean optimizing the project bottom lines. To compare technically feasible mining options, one must compare not only mining capital and operating costs but also corresponding processing costs, revenues from sales, infrastructure requirements, and environmental and social impacts.

The first requirement for successful mine planning is appropriate expertise and relevant experience. This experience must be in the type of deposit and mineral being mined. Past experience must include mine planning and development of achievable production schedules, equipment selection, and capital and operating cost estimation.

There is considerable information that the mining department must supply that directly or indirectly influences the financial bottom line, which include

- Initial mine development requirements and costs to be capitalized, including soil removal, access roads, pre-stripping, shaft sinking, major underground development, surface infrastructure;
- Total tonnage mined on a yearly basis (shorter time periods are used during the early years of the project and longer during the later years);
- Tonnage and quality of material mined and delivered to the processing plant, and corresponding costs;
- Tonnage and quality of material mined and delivered to the waste dump, and corresponding costs;
- Tonnage and quality of material mined and delivered to stockpiles and corresponding costs;
- Tonnage and quality of material recovered from stockpiles and delivered to the processing plant and corresponding costs;
- Other mining costs, including cost of waste dump and stockpile maintenance and environmental control, and mine overhead costs;
- Initial capital costs, including cost of fixed and mobile equipment, capitalized mine development, and infrastructure, offices, maintenance shop, change room;
- Yearly capital costs (sustaining capital and cost of future expansions, if any); and
- Mine closing costs.

Some information, such as capital and operating costs, influences the financial bottom line directly. Other information influences the financial bottom line indirectly. For example, the quantity and quality of material delivered to the processing plant influences the plant capital and operating costs and must be combined with plant recovery and market value of product sold to get revenues from sales.

MINING DECISIONS AND THE ENVIRONMENTAL BOTTOM LINE

The environmental bottom line is a measure of the impact that the project is likely to have on the natural, physical, and biological environment. In this section, only the environmental impact that can be directly attributed to that part of the project that is the responsibility of the mining engineering department is considered.

The project mining engineer must be experienced in taking environmental factors into account and knowledgeable of all applicable laws and regulations. The company's corporate objectives and social, environmental, and ethical values must be understood. The impact of mining activities that directly or indirectly influence the environmental bottom line can be classified as follows:

- Water. The impact of mining operations on surface and groundwater is one of the most important contributors to a project's environmental bottom line. To evaluate this impact, one must take into account the following:
  - Disruption of surface waters, including rivers, streams, lakes, ponds, irrigation channels, resulting from location of open pit, waste dumps, roads, offices, surface infrastructure, and so on.
  - Water consumption for mining operation, including dust control. The impact of water consumption by mining operations is significantly less than that from the processing plant but should not be ignored.

- Mine dewatering and water disposal. These activities are likely to impact both underground aquifers and surface waters. The expected quality of water discharged from pit or underground dewatering must be assessed, and the need for water treatment facilities must be evaluated. Costs must be included in the financial bottom line.
- Acid mine drainage and contaminants leaching from the pit, underground mine, waste dumps, and stockpiles are significant environmental and social risk factors. Facilities must be designed to minimize this risk, monitor quality, and treat waters before disposal.

- Soil erosion and contamination. Erosion control must be included as part of water management. Impacts can be in and around the mine, waste dumps, overburden piles, haul roads, stockpiles, reclamation areas, and so forth.
- Air pollution. Sources of air pollution include surface and underground mine operations, haulage roads, waste dumps, and stockpiles. Emissions from haulage trucks, mine equipment, and vehicular traffic contribute to air pollution. The cost of air pollution control, including capital and operating, must be assessed.
- Noise pollution. Sources include blasting, loading, haulage, ventilation, vehicular traffic, and so on. The impact must be taken into consideration in designing operations, selecting equipment, and locating facilities.
- Visual impact. The visual impact of infrastructure, waste dumps, and stockpiles can be minimized by appropriate location and design.
- Impact of blast vibrations, mine subsidence, and potential pit slope failure on roads, buildings, and other infrastructure. The corresponding social and environmental risk must be assessed and mitigated.
- Effect of mining activity on biological environment. Such impact may result from disruption of water sources, feeding grounds, and migration routes. Deforestation and site clearing, soil removal, water

and air contamination, noise, access roads, waste dumps all have an impact on the biological environment. Endangered species must be taken into account. The biological impact of mining must be assessed and mitigated.

- Mine closure. Mine closure must take into account current and future impacts. Surface reclamation includes access roads, waste dumps, maintenance facilities, exploration drill holes, headframes, offices, and other surface infrastructures. More controversial is pit closure, which must take into account future uses, long-term stability, and safety of access. Termination of dewatering may result in formation of a pit lake with long-term impact on underground water flows and water quality. Closure of underground mines requires analysis of long-term surface impact resulting from subsidence, as well as the likelihood of water discharge to the surface. The need for end-of-life land rehabilitation and consideration of future land use is likely to influence project design and operation. End-of-life reclamation costs can be decreased if concomitant reclamation is planned during operations.

MINING DECISIONS AND THE SOCIAL BOTTOM LINE

The social bottom line is a measure of the impact, positive or negative, that the project is likely to have on people. In this section, only the social impact that can be directly attributed to that part of the project that is the responsibility of the mining engineering department is considered.

The project mining engineer must be experienced in taking social factors into account and knowledgeable of all applicable laws and regulations. The company's corporate objectives and social, environmental, and ethical values must be understood. Mining activities that directly or indirectly influence the social bottom line will be different for each project. Each stakeholder is likely to have a different opinion concerning appropriate measurement of the social impact. The stakeholders' priorities can be summarized as follows:

- Shareholders, banks, and financial institutions. These stakeholders are primarily interested in the financial performance of the project. Their expectation is that financial performance will not be compromised because of failure to take appropriate consideration of social and environmental impacts. External audits of the three bottom lines are likely to be required. Financing organizations are likely to impose specific completion tests to be passed within specified time periods. Failing to pass these tests will result in financial penalties. Such tests may include physical, financial, environmental, and social targets. The mine must be designed to meet these targets with a high level of confidence.

- Business analysts. Their primary role being to inform investors, their opinion takes into account the interests of shareholders and financial institutions, with due consideration being given to social and environmental factors.
- International nongovernmental organizations (NGOs). With no financial interest in the project, NGOs view it as their mission to protect the environment and defend the interests of local communities and individuals who, in their view, may not be appropriately protected. Because the reach of NGOs is international, their opinions can have global repercussions. Early presentation of project objectives should be made to NGOs, which emphasize how environmental and social concerns will be taken into consideration. Some NGOs have the nonsustainable opinion that any form of mining should be stopped, thus denying the possibility of any constructive participation.
- Employees and their families. The workforce may be composed of local and expatriate employees who have different expectations concerning training, salaries, bonuses, and fringe benefits. Local culture, traditions, and religion may have an impact on work practices and productivity. All employees expect a safe work, family, and recreational environment. A rapid response team should be trained to react to mine accidents, need for evacuation, chemical spills, and so on. Appropriate medical facilities will be expected. When developing mine plans, the mining engineer should be aware of regulations concerning the percentage of nationals required to be trained and employed and how this percentage might influence costs and productivity.

- Customers and suppliers. Mines need millions of dollars of equipment in the form of drills, loaders, haul trucks, water trucks, light vehicles, and so forth. Mines also need consumables such as fuel, blasting supplies, electricity, and so on; and services for maintenance, and so on. To control costs and ensure quality, connections with suppliers should be initiated early and maintained throughout the operation. To the extent possible, supplies of goods and services should be obtained locally. Opportunities may be limited but should be considered and, in some cases, can be created by suitable financing and training. Accounting procedures should ensure that local suppliers receive fair treatment and are paid without undue delays.
- Local, regional, federal, and country governments. Mining engineers must be fully knowledgeable of all laws and regulations that apply in the country where operations are planned. These include, but are not limited to, work conditions, health and safety, and environmental compliance. The percentage of expatriates who can be employed is often regulated and required to decrease over time.
- Local communities. The causes of social conflict include lack of sufficient consultation and community engagement, lack of accurate information on mining impact, differing expectations of social and economic benefits, environmental concerns, and disputes over land use and economic compensation. Mining activities occupy large areas and require access to land, not only where the mine is to be located but also for processing plant, maintenance shops, laboratories, offices, leach pads, waste dumps, and so on. The mining engineer defines a significant part of these requirements and must take into account land ownership and usage to control the financial and social costs of displacing current owners. Artisanal and small-scale mining activities may already be taking place, legally or illegally, on the mine site; or may be initiated by local population when exploration is taking place, in the hope of getting compensation. Managing termination of such activities can be challenging. The role of government in removing artisanal miners must be ascertained, and removal procedures and compensations must be understood and accepted. Training and hiring the local population and buying products locally

will increase local acceptance. Mine operations and infrastructure can compromise accesses between communities or from communities to their traditional land or to market. Mitigation plans must be developed accordingly.

- Future generations. The design of mine operations, including equipment selection, should take into account expected emission of carbon dioxide and other greenhouse gasses that may contribute to global warming and climate change. A sustainable environment must be left behind after mine closure and reclamation.

RISK MANAGEMENT AND MINING ENGINEERING

The role of the mining engineer is to make recommendations concerning the technical and financial aspects of the expected mining facilities that influence the utility of the project, including the financial, social, and environmental bottom lines. To make these recommendations, the mining engineer relies on information supplied by others, as well as personal experience, knowledge, and judgment. Information that is used, of which the most important is arguably of geological nature, present some level of uncertainty, which is quite significant in early-stage development and decreases over time without ever becoming negligible. Competence of the mining engineer can add another level of uncertainty. The mining engineer needs to quantify the uncertainty concerning his or her estimates of realizable production rates, quantity and quality of material deliverable to process, estimates of capital and operating costs, and risk to the project that results from environmental and social impacts.

The role of the mining engineer was discussed extensively earlier in this chapter. Only some of the mining engineering decisions that can contribute to increased project risk are listed here:

- Developing mine plans without first discussing the resource model with the geologist. There are too many instances when resource models are "thrown over the fence" (transferred electronically) by modeler to mining engineer without any human interaction. Confidence that can be given to the different parts of the model is unknown, except for the resource being somewhat simplistically

SELECTIVITY AND RESOURCE MODEL

A scoping study was completed to determine the economic potential of a low-grade copper deposit. Initial assumptions included bench heights of 7 m, and blasthole spacing of 3.5 m. Ore control would consist of blasthole sampling, estimation of 3.5×3.5×7-m blocks by ordinary kriging, and design of ore control polygons that took into account mining face orientation and size of equipment.

During completion of the pre-feasibility study, the mining engineers considered a number of bench heights and blasthole spacing and determined that a 10-m bench height would be optimal, allowing the use of larger equipment and resulting in significantly lower mine operating costs. The mine geologists were also asked to investigate whether ore control costs could be reduced. The resource model showed that the high-grade zones were steeply dipping and highly continuous along strike. The decision was made to reduce ore control costs by using samples from widely spaced inclined reverse-circulation (RC) holes instead of close-spaced blastholes. The number of samples would be significantly reduced, resulting in savings expected to exceed the RC drilling costs. While some information would be lost, this was not expected to be significant. The quality of the samples would be improved, and the mining engineers would get information over two to three benches instead of only one, thus improving short-term planning.

When operations started, it became immediately evident that the expected head grade could not be reached. The tonnage of mill feed delineated by the new ore control method was higher than expected, but the grade was much lower. Analysis of the results showed the following:

- The resource model was based on 7-m bench heights and grade information obtained from close-spaced blastholes. Changing the bench height to 10 m, using larger equipment, and modifying the ore control method resulted in significantly lower selectivity than that assumed in the resource model.
- A new resource model should have been developed that took into account the change in selectivity.
- The mine plan should have been updated using the new resource model. Most probably significant changes would have been justified, including use of a lower cutoff grade and increased mine capacity, and commensurate changes to the processing plant. Changes to mine and mill capital and operating costs would have resulted.
- An updated project feasibility study may or may not have confirmed that the higher bench height was indeed economically justified.

classified as measured, indicated, or inferred. The result is that geologic uncertainty is likely to be mostly underappreciated or ignored in mine design.

- Locating surface and underground facilities with insufficient knowledge of deposit characteristics in areas surrounding the main deposit. Location may have to be changed when additional sterilization drilling has been completed.
- Comparing a number of mining options (open pit versus underground; 7-meter bench versus 10-meter bench; ore control using blastholes versus reverse-circulation holes) using a single resource model. Most resource models are based on specific mining and processing assumptions and may not be valid if these assumptions are changed.
- Underestimating time and cost of production ramp up. In open-pit mines, production start-up often requires pre-stripping of waste, near-surface material that cannot be processed, or material with low metallurgical recovery. Areas to be mined first must be characterized with significantly more topographical, geological, and geometallurgical information than those mined later in the project life.
- Underestimating time and cost of construction of mine infrastructure in underground mines. If properly planned, the time and cost of acquiring appropriate geotechnical information ahead of shaft sinking and underground drifting is significantly less than the delays likely to result from unforeseen geological features encountered during construction.
- Overestimating productivity resulting from impractical mine design (ramps too steep or too narrow, not enough space for equipment in the loading areas, etc.) and optimistic assumptions concerning equipment utilization (assuming no truck delays at the shovels, no waiting time in ore dumping areas).
- Locating surface facilities without appropriate soil and subsoil information, resulting in unstable foundations for waste dumps, stockpiles, primary crusher, offices, and laboratories.

- Insufficient environmental data to locate surface facilities, including buildings, stockpiles, and waste dumps.
- Underestimating mobile equipment requirements (trucks, shovels, water trucks, dozers, explosives delivery trucks, light vehicles, etc.) and corresponding capital and operating costs.
- Underestimating the need for maintenance facilities for both mine equipment and light vehicles and associated capital and operating costs.
- Underestimating site overhead, office space, computer hardware, and software requirements, need for data and voice communications.
- Underestimating ore control requirements, including staffing, vehicles, sampling, laboratory facilities, and database management.
- No allowance made for health and safety training facilities and emergency services.
- Underestimating time and cost of construction of surface facilities, offices, roads, waste dump site preparation, water management facilities, environmental controls, and so on.
- Underestimating operating costs, unrealistic assumptions concerning equipment life and productivity, and underestimating sustaining capital.

This partial list is sufficient to give an idea of the many risk factors that can influence project evaluation. The mining risk can be mitigated by nominating a senior mining engineer with practical relevant experience in both mine planning and mine operations, either to be part of the study team or to periodically review assumptions, designs, and costs. Reviews should take place periodically, when the study is taking place, and not be limited to the final design. Reviews completed at the end of a study are of limited effectiveness. At this late stage, final decisions have been made, expectations are set, everyone is in a hurry to move forward, and there can be considerable pressure toward ignoring defects, or even fatal flaws, thus significantly increasing the risk of project underperformance or failure.

CHAPTER 15

METALLURGY AND PROCESS ENGINEERING

Since metalliferous ores differ greatly among themselves, first as to the metals which they contain, then as to the quantity of the metal which is in them, and then by the fact that some are rapidly melted by fire and others slowly, there are, therefore, many methods of smelting. Constant practice has taught the smelters by which of these methods they can obtain the most metal from any ore. Moreover, while sometimes there are many methods of smelting the same ore, by which an equal weight of metal is melted out, yet one is done at a greater cost and labour than the others.

Georgius Agricola, *De Re Metallica*, 1556

The role of the process engineer is to define the most efficient way to extract a salable product from the material that the mine can deliver. Processing efficiency must take into account input constraints imposed by the properties of the material to be delivered by the mine and how these properties are expected to change during the mine life. Output constraints might be imposed by the market, which may influence quantity, quality, and value of salable product. Capital and operating costs are important determinants of process efficiency.

RELATIONSHIP BETWEEN PROCESSING AND OTHER TECHNICAL DEPARTMENTS

The metallurgical processes that are applicable to a specific project are determined by the deposit characteristics and available technology. Process risk

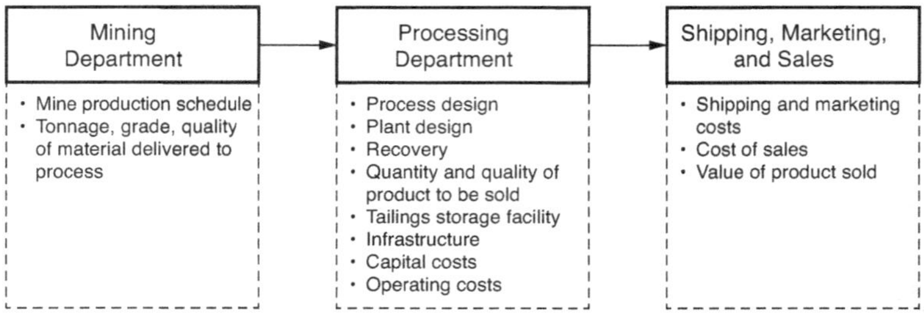

FIGURE 15.1 Linear (oversimplified) decision-making in process engineering

increases when the geology presents unusual properties and when a new metallurgical process must be developed. The process design must satisfy social, environmental, legal, and safety regulations, as well as take into account company objectives and values. The three bottom lines—financial, social, and environmental—must be taken into account when comparing processing options.

Initially, when only limited information is available concerning the deposit, the approach to decision-making in process engineering can be considered as linear. The mine supplies material to the processing plant on a specified schedule. The processing plant extracts a valuable product from the material it receives and disposes of rejects in an environmentally sound manner. The valuable product is delivered to market. The main objective is to maximize extraction of valuable product while taking into account all costs. This can be represented as shown in Figure 15.1.

In actuality, the decision-making process is significantly more complex. Metallurgical information is needed in the early stages of exploration, when a mineralized deposit has been discovered but there is not sufficient information to develop a meaningful production schedule. To be defined as an inferred mineral resource, a deposit must show reasonable prospect for eventual economic extraction of a salable product. Early metallurgical tests are needed to determine extraction processes that might be technically and economically feasible. A number of processes may be applicable to different parts of the deposit. Metallurgical tests are also needed to guide exploration as to which information should be collected to assist future decisions. A meaningful geometallurgical model cannot be developed without prior understanding of likely metallurgical process requirements. This is represented by Figure 15.2.

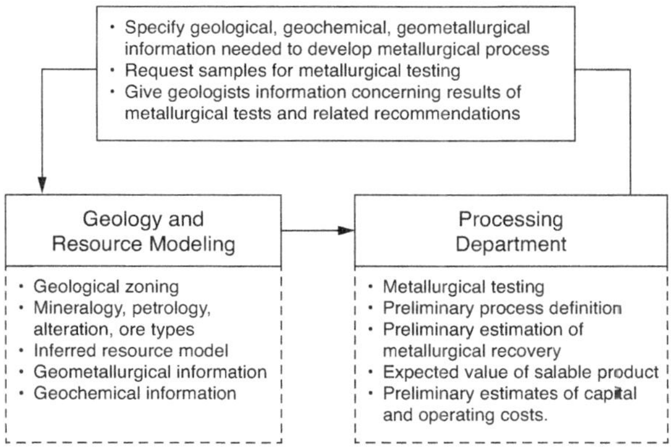

FIGURE 15.2 Role of processing department during exploration and early project evaluation

When information improves, production schedules of increasing reliability are developed. Such schedules require estimation of cutoff grades, which cannot be determined without estimates of processing costs and process recoveries. Achievable mine production schedules depend on geology, applicable mining method, and realistic mine design. The capacity of the processing plant is first determined by the quantity and quality of material that the mine can deliver over time. Optimization of production schedules demands balancing mine and process capital and operating costs as well as revenues from sales. This is better represented by Figure 15.3, which shows the wide variety of information that must be available to develop production schedules, to estimate process capital and operating costs, and to determine revenues from sales. There is a need for significant information feedback among geology, mining engineering, process engineering, infrastructure engineering, and shipping, marketing, and sales departments. The linear process of Figure 15.1 does not apply except very early in the evaluation of exploration projects.

The mine production schedule is the primary input to the processing plant. This production schedule must specify quantity and quality of material sent to the plant, including information needed to determine plant capacity, throughput, recovery, and quantity and quality of salable product. The presence of contaminants in the feed must be known, as well as how these contaminants will affect the product market value or the environmental properties of rejects. The metallurgist must estimate plant capital and operating costs, as

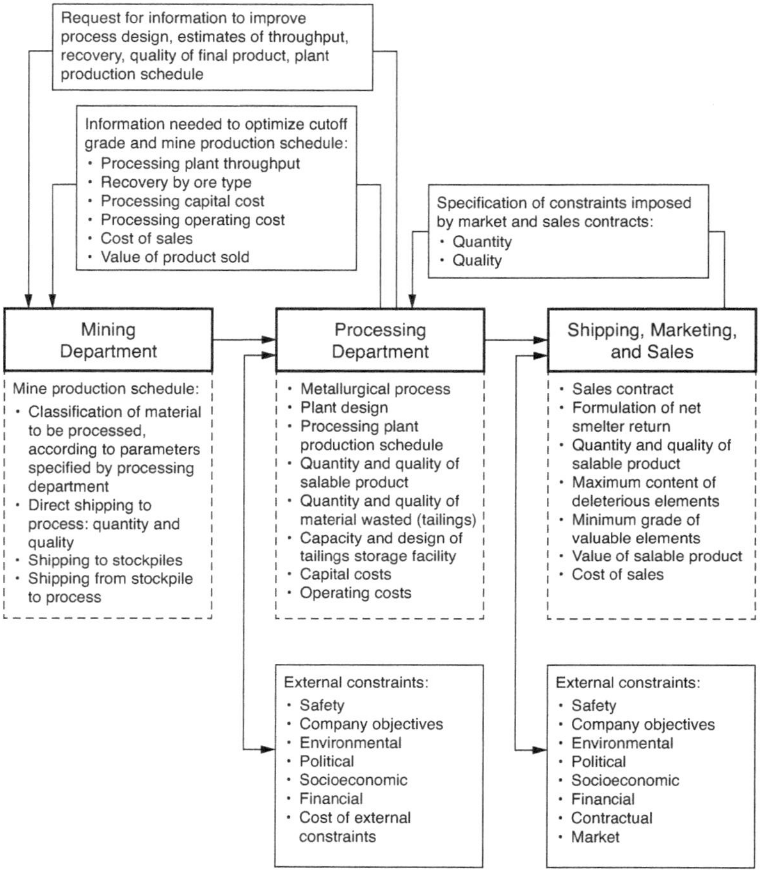

FIGURE 15.3 Complex (realistic) decision-making in process engineering

well as recovery and value of product sold to make it possible for the mining engineer to optimize mine plans and production schedules. Marketing requirements such as constraints on quantity and quality of iron ore, industrial minerals, or copper concentrate that the market will accept have a direct impact on process design, plant capacity, and the mine production schedule.

ROLE OF THE PROCESSING DEPARTMENT

A competent person must determine which metallurgical process should be used and take responsibility for plant design, capital and operating cost estimates, process recovery, and value of product sold. The competent person must estimate the cash flow that is attributable to the processing activities. This cash flow has a negative component, including capital and operating costs, and a positive component resulting from sales. The competent person must determine when sufficient information is available to declare a mineral resource or a mineral reserve.

The design of the processing plant must take into account the quantity and quality of material that can be delivered over time by the mine. These quantities and qualities are a function of the cutoff grade, which is estimated from expected mining and processing costs, process recovery, and value of salable product. Costs, and therefore cutoff grades, are dependent on plant size and processing rate. Planning is, by necessity, an iterative process. Preliminary assumptions are made, a production schedule is developed, the original assumptions are refined, and a new schedule is developed with increasingly reliable estimates of both input and output variables. The process is repeated until sufficient confidence has been reached and the cost of further risk reduction— including reduction in probability and consequences of unfavorable events—is no longer considered justified.

During the early stages of project development, metallurgists must work closely with geologists to ensure that there is mutual understanding of the information that is needed to determine whether a salable product can be extracted. The objective should be to collect the necessary geometallurgical information in a timely fashion. Technical requirements and financial constraints must be taken into account when deciding what information is needed.

In early-stage exploration, when there is no assurance that an economic project will be developed, few samples may be needed. Geometallurgical testing is expensive and is only justified on a limited basis to estimate the order of magnitude of tonnage and grade of potentially economic mineralization. However, waiting until the project is sufficiently advanced to acquire geometallurgical information can result in duplication of work, re-drilling areas previously drilled, cost increases, and project delays.

When the project moves from exploration to pre-feasibility and feasibility study, the geological understanding of the deposit keeps improving and so must the information needed to define reliable geometallurgical models. The role of the process engineer becomes increasingly significant and is likely to include the following tasks:

- Review all applicable rules, regulations, and company objectives and values, including those related to safety, environmental compliance, and social responsibility.
- Review project information with geologists and mining engineers. Particular attention should be given to the geological properties of the deposit most likely to influence decisions concerning metallurgical characterization.
- Review metallurgical tests completed to date. Determine which processing methods are technically feasible and whether additional information is needed to improve process design. This is likely to require additional testing of available core samples, new large-diameter core samples, or bulk samples obtained from surface or underground workings.
- Use a preliminary production schedule suggested by the mining engineers to estimate processing plant capacity, capital and operating costs, and recovery and unit value of product sold.
- Mine production schedules are based on preliminary estimates of processing technical and financial parameters. As these estimates get refined, so do cutoff grades and production schedules.
- This process of optimization is repeated to take into account the availability of new technical and financial information concerning geology, deposit model, mine production schedule, and costs.
- Cost estimates must include processing plant, leach pads, tailings facilities, laboratories, pipelines, conveyors, offices, and other necessary infrastructure.
- Location of all facilities must take into account geography, surface and underground geology, and mine location including any possible future expansion. Also to be considered is likely social and environmental impact, including the need for relocation of previous landowners, changes in water flows, impact on fauna and flora, and so forth.

- Design changes are likely when more information becomes available and the studies progress from scoping to pre-feasibility and feasibility. A number of process options should be considered during the pre-feasibility study, with one option being retained at the end of the study. This option is that which will form the basis for a feasibility study.
- Eventually, a detailed design of the processing plant and ancillary facilities will be required. This is most commonly done in cooperation with an engineering, procurement, and construction (EPC) company. The role of the project metallurgists and process engineers is to ensure clear scope definition and quality control. Representatives of the mining company who will be responsible for plant operation must be heavily involved in this feasibility study, as they eventually will have to accept full responsibility for technical and financial performance.
- The scope of work must be well defined before construction starts. The temptation may be to start construction early, with limited planning, thus increasing the apparent project net present value. However, insufficient planning and premature start of construction is likely to result in scope changes, either during construction or shortly after. The right level of planning is that which takes into account the probability that scope changes will be required and the delays and costs that will result.

The choice of metallurgical process and the decisions made concerning plant design directly influence the project utility. The ways in which decisions made by the process engineer influence the three bottom lines are summarized in the rest of this chapter.

PROCESSING DECISIONS AND THE FINANCIAL BOTTOM LINE

The cash flow generated by the processing department has both a negative and a positive component. The negative part is measured by capital and operating costs. The positive part is determined by the value of the material produced and sold as a result of processing the material received from the mine. The processing department faces two conflicting objectives: minimizing costs while maximizing revenues from sales. Optimization is subject to constraints imposed by other parts of the operation, especially mining.

For reliable decision-making, the processing department must have demonstrated proficiencies in the estimation of all related technical and financial aspects of a new project. Information that must be supplied, which influences the financial bottom line in a significant way, includes the following.

- Capital costs:
  - Site preparation for processing plant, leach pads, laboratories, tailings storage facilities, access roads, offices, and so on
  - Sources of material for tailings storage facility, and cost
  - Sources of water for plant operation and cost of water supply
  - Infrastructure for disposal of excess water
  - Power supply—power plant, backup generators, power lines
  - Initial capital cost, including engineering and construction of processing plant, maintenance facilities, laboratories, leach pads, tailings storage facility, infrastructure, offices, mobile equipment, and so on
  - Capitalized start-up costs
  - Sustaining capital and cost of future expansions, if planned
  - Cost of environmental control facilities
- Operating costs:
  - Staffing (plant operation and maintenance, management, metallurgical engineering, laboratory, safety, training)
  - Process operating costs
  - Maintenance costs
  - Laboratory operations and quality control

  - Consumables (grinding media, chemicals, electricity, water, others as required for specific circuit)
  - Overhead
- Revenues from sales:
  - Tonnage and quality of material processed on a yearly basis
  - Metallurgical recoveries, quantity and quality of product available for sale
  - Cost of sales
  - Unit value of product sold— metal price, net smelter return
  - Revenues from sales, estimated on a yearly basis
- Closure costs:
  - End-of-life one-time closure costs (processing plant, leach pads, tailings facilities, laboratories, dams, roads, buildings, etc.)
  - Cost in perpetuity (to control water contamination from tailings facilities, and monitor long-term stability of tailings dams)

PROCESSING DECISIONS AND THE ENVIRONMENTAL BOTTOM LINE

The environmental impact is defined as that which affects the physical and natural environment. Metallurgists and process engineers must be experienced in taking environmental factors into account and knowledgeable of all applicable laws and regulations. The company's corporate objectives and social, environmental, and ethical values must be understood. Information that is needed to evaluate the environmental bottom line includes the following.

- Water. The impact of processing operations on surface and underground water is one of the most important contributors to the environmental bottom line. Included should be consideration of the following:

- Processing plant water requirements: In most mining operations, the processing plant is the most significant water consumer. Evaluation of the potential environmental impact starts with correct estimation of water requirements and development of plant and site water balance.
- Sources of water: There are a number of possible sources of water, including mine dewatering, major rivers and lakes, underground aquifers, purposely built retention dams, and, in some instances, the ocean. Each possible source of water must be assessed for the following:
    - Suitability—Is there enough water of sufficient quality to satisfy the plant needs?
    - Availability—Who owns these waters? Can water rights be acquired?
    - What will be the environmental and social impact of consuming these waters? How will this impact be mitigated?
- Water recycling: Recycling process water may be an effective way to minimize environmental impact. Such recycling must be planned during project design because it can have a significant impact on capital and operating costs, especially as they relate to tails handling and construction of tailings dam facilities.
- Leach pads: The environmental impact of leach pads can be surficial and underground. Surface use of land has an impact on plant and animal life. Care should be taken to avoid water contamination from pad failure or overflow from ponds during extreme weather events.
- Tailings disposals: Leakage from tailings can pollute surface or underground waters and needs be monitored. Tailings facilities must be designed to minimize the risk of such leakage. Use of paste tailings can be an effective but expensive way to reduce the risk of water pollution and dam failure, increase recycling of process waters, and reduce reclamation and closure costs. Ocean disposal of tailings remains controversial because consequences are not fully understood.

- Air pollution: The impact of air pollution is easily seen today around old processing facilities, some of which are still operating. It touches human health and the animal and plant environment. It results in acid rain, and soil and water contamination. The need for emission controls to ensure acceptable air quality varies depending on the composition of the material being processed and the process being used. Such controls can represent a significant part of the capital and operating costs. Power generation contributes to air pollution. Crushing and conveying can be significant sources of air pollution. Wind-blown dust from stockpiles, leach pads, and tailings storage facilities must be controlled. Suitable and reliable air quality control systems must be designed and maintained, and their effectiveness monitored. Emergency plans must be in place for rapid response if pollution controls fail.
- Noise pollution: Processing facilities can be very noisy operations and their design and location must take into account the human and biological environment that will be affected by this noise.
- Biological environment: Leach pads and tailings facilities can impact very large areas. The location of all process facilities must take into account the preexisting biological environment, including vegetation, surface waters, migration patterns, and so on.
- Tailings dam failures: History is replete with catastrophic failures of tailings storage facilities with major environmental consequences, such as contamination of soils and aquifers, and destruction of wildlife and vegetation. Tailings storage facilities should be built to the highest standards to minimize the likelihood of such failure. A full assessment should be made of the consequences if failure does occur, including financial, social, and environmental impacts, as well as consequences concerning the company's future license to operate.

PROCESSING DECISIONS AND THE SOCIAL BOTTOM LINE

The social bottom line is a measure of the impact, positive or negative, that the project is likely to have on people. The process engineer must be experienced in taking social factors into account, and knowledgeable of all applicable laws and regulations. The company's corporate objectives and social, environmental, and ethical values must be understood. Processing activities that directly or indirectly influence the social bottom line will be different for each project. Each stakeholder is likely to have a different opinion concerning the significance of the social impact of a processing plant. The stakeholders to be considered include

- Shareholders, banks, and financial institutions;
- Business analysts;
- International nongovernmental organizations (NGOs);
- Employees and their families;
- Customers and suppliers;
- Local, regional, federal, or country governments;
- Local communities; and
- Future generations.

Expectations concerning each stakeholder, and how to mitigate these expectations, were presented in Chapters 11 and 14. While Chapter 14 refers specifically to mining engineering, most of the observations made in this chapter with respect to the social impact of operations are also applicable to process engineering. The stakeholders' priorities can be summarized as follows:

- Shareholders, banks, and financial institutions. These stakeholders are primarily interested in the financial performance of the project, which is dependent on process capital and operating costs, and predicted sales. Financing organizations are likely to impose specific financial and technical completion tests, which the processing plant must pass to avoid significant penalties.
- Business analysts. Their primary role being to inform investors, their opinion takes into account the interests of shareholders and financial institutions, with due consideration being given to social and environmental factors.

- International nongovernmental organizations. NGOs view their mission to protect the environment and defend the interests of local communities. Early presentation of project objectives should be made to NGOs, which emphasize how environmental and social concerns will be taken into consideration.
- Employees and their families. The workforce may be composed of local and expatriate employees who have different expectations concerning training, salaries, bonuses, and fringe benefits. Local culture, traditions, and religion may have an impact on work practices and productivity. All employees expect a safe work, family, and recreational environment.
- Customers and suppliers. The processing plant is likely to be the most significant part of the total capital cost of a new mining operation. The same can be said of the operating cost. To control these costs and ensure quality, connections with suppliers should be initiated early and maintained throughout the operation.
- Local, regional, federal, and country governments. Process engineers must be fully knowledgeable of all laws and regulations that apply in the country where operations are planned.
- Local communities. The causes of social conflict include lack of sufficient consultation and community engagement, lack of accurate information on mining impact, differing expectations of social and economic benefits, environmental concerns, and disputes over land use and economic compensation. Process facilities occupy large areas and require access to land for the plant, maintenance shops, laboratories, offices, leach pads, stockpiles, and so on. Location and design of these facilities must take into account land ownership and usage to control the financial and social cost of displacing current owners.
- Future generations. Appropriate controls of carbon dioxide and other greenhouse gasses must be put in place. A sustainable environment must be left after plant closure and reclamation.

PROCESS ENGINEERING AND GEOMETALLURGICAL CHARACTERIZATION

The need for effective communication between geologist, metallurgist, and process engineer was discussed earlier, in the Chapter 12 sidebars. Listed here are other examples where lack of effective communication can result in significant cost increases:

- Multi-element geochemical tests were completed by the geologist but not reviewed with the metallurgist. When the flotation plant was commissioned, unexpected high concentrations of contaminants were found in the concentrate, which would have been predicted if a proper analysis of the available geochemical data had been completed. The high level of contaminants was outside that agreed in the smelter contracts with ensuing costly penalties.
- When evaluating the economic feasibility of an underground mine, metallurgical tests were performed on mineralized material only. The narrow-vein mining method added significant footwall and hanging-wall waste dilution to the material sent to process. The metallurgical properties of this waste had not been tested. Initial performance of the processing plant was significantly lower than expected. Plant modifications had to be made. The problem could have been avoided by appropriate early communication between geologist, mining engineer, and metallurgist.
- In the geologic model of a porphyry copper deposit, a subhorizontal surface was used to define the boundary between oxide and sulfide zones. Multiple vertical fault zones had been recognized and were included in the structural model. Also recognized by the geologists was that these faults brought oxidation deep inside the mostly sulfide domains. However, this complexity was ignored in the geometallurgical model, the tonnage of oxide material in faults being relatively small. Failure to include this information had significant repercussions: first, when oxide material was mixed with sulfide sent to the flotation plant, resulting in substantial loss of recovery, and then when mining and ore control procedures had to be modified to separate oxidized fault material from surrounding sulfide mineralization.

RISK MANAGEMENT AND PROCESS ENGINEERING

When evaluating a project, the role of the process engineer is to make recommendations concerning the technical and financial aspects of the processing facilities that influence the financial, social, and environmental bottom lines.

To make these recommendations, the process engineer relies on information supplied by others, as well as personal experience, knowledge, and judgment. Information that is used, of which the most important is the quantity and quality of material that the mine will deliver to the processing plant, present some level of uncertainty. This uncertainty is quite significant during the early stages of project evaluation and decreases over time without ever becoming negligible. Competence of the process engineer can add another level of uncertainty. The process engineer needs to quantify the uncertainty concerning his or her estimates of realizable processing rates and recoveries, value of product delivered to market, estimates of capital and operating costs, and environmental and social impacts.

The role of the process engineer was discussed extensively earlier in this chapter. Only some of the process engineering decisions that can contribute to increased project risk are listed here:

- Requesting samples for metallurgical testing without first reviewing the deposit geological characteristics with the project geologist. Depending on his or her experience, a representative sample as defined by the geologist may not be what a metallurgist needs to characterize the deposit.
- Not enough consideration of where and how many samples should be taken, how they should be prepared, and whether they should represent blended material or not. The answer is primarily a function of geology but also of mining sequence and processing rate.
- Designing a process that only takes into account the results obtained from testing a single large sample in a pilot plant. While pilot plants can supply high-value information, the likely variability of plant feed over time must be understood before finalizing a design.
- Underestimating time and cost of plant design and construction. The plant owner must ensure that appropriate quality control and cost accounting procedures are in place to supervise the work completed by the EPC firm responsible for construction.
- Underestimating time and cost of reaching commercial operation. Start-up is when plant defects become most apparent. The ore

presented to the plant is not likely to be representative of "typical" material used to design the plant. Unless start-up conditions are understood and expectations are properly set, plant performance is likely to be seen as disappointing. This is particularly true in open-pit mines in which geometallurgical characterization of near-surface material needs particular attention.

- Inappropriate design and location of tailings storage facilities resulting from insufficient consideration given to environmental factors, quality of material used for dam construction, water treatment facilities, and so on.
- Underestimating site overhead, office space, computer hardware and software requirements, and need for data and voice communications.
- Underestimating laboratory requirements, including staffing, vehicles, equipment, quality assurance and quality control, and database management.
- No allowance made for health and safety training facilities and emergency services.
- Underestimating operating costs, maintenance costs, and sustaining capital.

This partial list is sufficient to give an idea of the many risk factors that can influence project evaluation. The metallurgical risk can be mitigated by nominating a senior metallurgist and process engineer who has practical relevant experience in process planning, plant design, and plant operations to be part of the study team or to periodically review assumptions, designs, performance, and cost estimates. Reviews should take place periodically, when the study is taking place, and not be limited to the final design. Reviews completed at the end of a study are of limited effectiveness. At this late stage, final decisions have been made, expectations are set, everyone is in a hurry to move forward, and there can be considerable pressure toward ignoring defects or even fatal flaws, thus significantly increasing the risk of project underperformance or failure.

CHAPTER 16

INFRASTRUCTURE

The miner should not contemplate mining without considering whether the place is covered with trees or is bare. If it is a wooded place, he who digs there has this advantage, besides others, that there will be abundant supply of wood for his underground timbering, his machinery, buildings, smelting, and other necessities.... The miner should consider the locality, as to whether it has a perpetual supply of running water, or whether it is always devoid of water except when a torrent supplied by rains flows down from the summits of the mountains....The miner also should consider whether the roads from the neighbouring regions to the mines are good or bad, short or long.

Georgius Agricola, *De Re Metallica*, 1556

In the previous chapters, the three fundamental technical components of a mining project were discussed: exploration, mining, and processing. No mining company can hope to be successful without exceptional strengths in these areas. The responsibilities of mining engineers and metallurgists include design and pricing of the facilities that are needed to bring the mine and processing plant to commercial operation. This infrastructure is not all-encompassing. Additional infrastructure is required to support the project as a whole, which has technical, financial, environmental, and social consequences. Failure to adequately evaluate these requirements can significantly increase the risk that the project will not perform as planned.

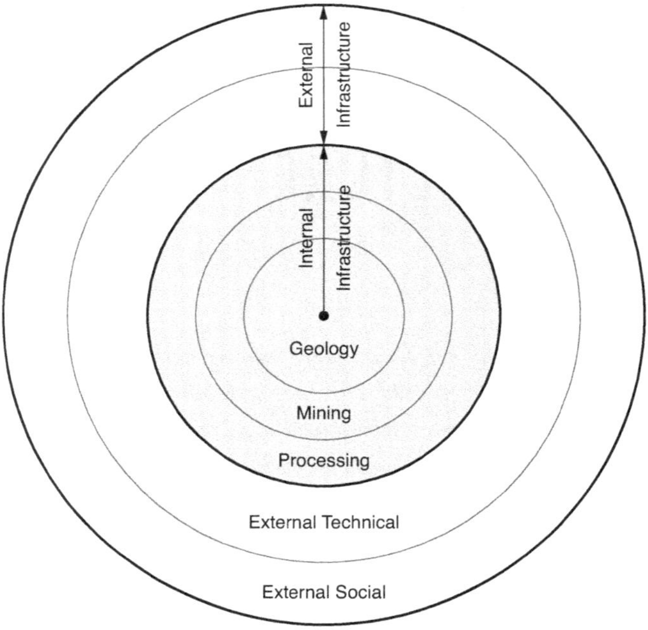

FIGURE 16.1 Increasing circles of infrastructure requirements

INTERNAL AND EXTERNAL INFRASTRUCTURE

Infrastructure can be classified as *internal*, which is located within the perimeter of the mining operation, and *external*, which is located outside this perimeter. Internal infrastructure is that needed to support mine-site operations, including the geologic, mining, and processing activities. External infrastructure is best subdivided as "technical" and "social." This infrastructure is needed to ensure that not only the technical aspect of the project operates effectively but also that a socially and environmentally stable setting is maintained.

Concentric circles can be used to graphically represent the different types of infrastructure (Figure 16.1). The radius of each circle reflects the distance between the mineral deposit at the center and the physical location of the infrastructure needed to develop this deposit. The geologic part of internal infrastructure is at the center, surrounded by mining and processing. The technical part of external infrastructure is that which is needed to support both the project internal needs and the external social and environmental requirements. The outmost circle represents the external social infrastructure.

Social infrastructure refers to facilities, services, and installations that are needed for the functioning of the human environment affected by the mining operation. Included may be schools, recreational facilities, police, and so on. One might expect such infrastructure to be supplied and maintained by local, regional, or country governmental entities. However, such authorities may not have the necessary expertise or resources, especially in remote areas and developing countries. Although the preexisting social infrastructure may be adequate before mining, it may require significant improvement if it is to support the changes that mining will bring.

INTERNAL INFRASTRUCTURE

The success of a mining project is highly dependent on completion, on time and within budget, of the infrastructure needed before a positive cash flow can be generated. Part of the infrastructure is likely to be built by the mining company, such as access roads, preparation of waste dumpsites, or pre-stripping of an open pit. Other parts will be contracted to an engineering, procurement, and construction (EPC) company. This is usually the case for construction of the processing plant or for sinking shafts for an underground mine. Whether the mining company or an EPC company builds the infrastructure, the ultimate responsibility for quality, timing, and cost of completion resides with the mining company.

Internal infrastructure needs were discussed in earlier chapters and can be summarized as follows:

- Mine initial capital projects, including pre-stripping for open-pit mines, and shaft sinking and primary accesses for underground mines
- Haulage roads
- Sites for waste dumps and stockpiles
- Processing plant and leach pads

- Initial tailings storage facilities
- Laboratories
- Pipelines, conveyor belts
- Storage for consumables (explosives, fuel, oil, gas, etc.)
- Offices
- Training and safety facilities
- Emergency medical facilities
- Security facilities
- Road network for mine, processing plant, offices, tailings storage facilities, waste dumps, and so forth
- High- and low-voltage electrical power distribution network and emergency electricity generation facilities (diesel generators)
- Fuel (diesel and gasoline, natural gas, etc.) storage and distribution network
- Data and voice communications network (land, wireless, satellite, radio)
- Water dams, water supply
- Flood control dams
- Housing if on-site (permanent housing, recreational facilities, food supply, camp facilities)
- Sewage treatment facilities, water treatment plant
- Heliport, airport, as needed for access and emergency services
- Electric power station if reliable power is not available from national network

There is uncertainty in defining what to build (technical uncertainty) and what it will cost to build it (financial uncertainty). Cost estimates should take into account two components:

1. "Overnight" construction costs, which is the cost of a construction project if no interest was incurred during construction
2. Lead time, which is the time between initiation and completion of construction project, between beginning of project and appearance of results

Insufficient planning is one of the main reasons why projects fail to perform as planned. Planning requires adequate information and time to analyze this information. Many projects fail because the decision to go ahead with mine development was made prematurely. The sources of uncertainty are often cumulative. If the geological and geometallurgical properties of the deposit are poorly defined, so is the metallurgical process and the size of the plant that will be needed. An engineering company can be very effective in building a plant according to owner-specified requirements, but these requirements may not be those that suit the deposit. Scope changes resulting from original designs based on inadequate information are one of the main sources of cost overruns.

Given the high cost of infrastructure, one cannot help wondering why decisions to go ahead with mine development are often made without adequate information. The main reason appears to be an overestimation of the opportunity cost of postponing a project, and an underestimation of the risk taken when building a project with inadequate information. Consider an early-stage project of which the expected net present value (NPV) is \$1 billion (\$1B) using an 8% discount rate. The cost of postponing the project by two years, measured by the resulting decrease in NPV, is estimated as follows:

$$\$1\text{B} \times [1 - 1/(1{+}8\%)^2] = \$143{,}000{,}000$$

This opportunity cost is significant, but is it meaningful if the estimated infrastructure capital cost is also \$1B? An accepted range of accuracy for a feasibility study based on reliable information is 15%, which in this case would amount to \$150,000,000. If this capital cost estimate is based on unreliable information, such as poorly defined geological and geometallurgical deposit properties, lack of soil and subsoil characterization, uncertainty concerning land access, and so on, the downward risk will be significantly higher, say, 25% or \$250,000,000. The cost of postponing the project is considerably less than the cost of going ahead without adequate information.

EXTERNAL INFRASTRUCTURE

Technical Requirements

Infrastructure external to the mine site is needed to obtain consumables, goods, and services, and to deliver products to market. Such infrastructure may have to be built by the company or improvements may have to be made to existing facilities. Included might be the following:

- Vehicular communication network, including roadways, tunnels, and bridges, which may have to be newly built or improved to accommodate the increased capacity that the mine requires for transport of equipment, consumables, and personnel
- Rail transportation network or pipeline as needed to ship commodity (coal, iron ore) or processed mineral (concentrate) to market
- Shipping docks, laydown yard, and stockpile areas
- For isolated mine sites, air transportation, including airport or heliport and adjoining support facilities
- Electrical network to link the mine facilities to the national grid, if such a grid can support the additional demand, both routine and peak
- Power station and backup power supply if national grid does not have sufficient capacity or supply is unreliable
- Pipelines and related facilities to supply energy in the form of natural gas or liquid fuel
- Dams, pumping stations, pipelines, and water treatment plants to supply water to the mining facility and dispose of excess or contaminated waters in an environmentally friendly manner
- Telecommunications network—voice and data, ground, cellular or satellite

The capital and operating costs of this infrastructure must be included in the feasibility study. Although the impact of the technical infrastructure is primarily on the financial bottom line, it is also significant on the environmental and social bottom lines. To estimate these impacts, one must follow these steps:

1. Meet with local and central government representatives to keep them informed of the company's intentions, discuss project characteristics, and understand and manage expectations. Government cooperation is needed to understand power structures, local traditions, concerns, hopes, and needs.

2. Evaluate the infrastructure needs. These needs vary considerably from one operation to the other. They are a function of the size of the operation, the mining and processing methods, and the type of product to be sold. A large open-pit mine, which is designed to use 300-ton trucks, will have infrastructure requirements different from a small underground mine. Width and load rating of bridges and roads needed to deliver equipment to the site will not be the same. A higher traffic volume must be accommodated to acquire goods and services and to ship product to market. A mine producing gold doré might consider an airport to send its product safely to the refinery, while a copper mine selling concentrate may require pipelines and port facilities. A fly-in/fly-out operation demands specific infrastructure for personnel transportation, housing, food supply, and entertainment, which differ from those needed for permanent on-site employees.

3. Evaluate the existing infrastructure and the extent to which it matches the company's requirements. This evaluation is likely to start with review of official documents concerning existing roads, railroads, harbors, electrical supply, fuel and gas distribution networks, and so forth. Field reviews will be necessary to confirm the reliability of this documented information.

4. Evaluate the need to improve existing infrastructure or build new infrastructure. Evaluate the time required to upgrade the infrastructure, and the corresponding capital and operating costs. Depending on the complexity of the project, this evaluation might be completed by the mining company or by a contracted engineering, procurement, construction, and (sometimes) management company.

5. Gather the information needed to estimate total infrastructure costs, which includes topographic and geologic maps (soil, subsoil, and bedrock), as well as environmental and social information.
6. Infrastructure design must consider the probability of extreme natural disasters such as floods, avalanches, droughts, landslides, and earthquakes. Historical statistics must be modified to take into account global warming and changes in weather patterns.
7. Available information is often limited and may not be reliable when the required infrastructure extends over large remote areas, as may be the case with pipelines, rail facilities, or roads. Accurate information must be obtained if sound decisions are to be made. Failure to do so is likely to result in significant errors in estimation of cost and time needed to construct the necessary infrastructure.
8. The total cost of building this infrastructure must include the cost of obtaining right-of-ways and of mitigating social and environmental impacts that will result from construction and operation of the built facilities. Early communication should be planned with those communities that will be impacted in an effort to minimize delays and construction interruptions because of opposition to the project.
9. Also included must be the closing costs, which in some situations may include transfer of facilities to others, including governments, or returning areas to a similar preexisting state.
10. An understanding must be reached concerning who is going to pay for which part of the new or improved infrastructure: Will it be the central or local governments, possibly backed by complex arrangements with the World Bank or regional development banks, the public utility companies, independent investors, or the mining company? Tax agreements, national or international sources of funding, sharing of cost with customers, and other sources of creative financing are some of the sources of funding to be considered.
11. During the pre-feasibility study, different technical infrastructure options should be considered, taking into account financial, environmental, and social costs. A single option must be retained before the project progresses to feasibility study.

Social Requirements

The mine's social impact starts with construction. Indeed, the construction period is often that which has the most significant, albeit temporary, impact because of the very high level of activity taking place and number of imported workers moving into the community. Once the mine is in operation, the social impact is different, more stable, and of longer duration. The impact of mine closure must also be considered.

Social infrastructure refers to those institutions that are needed to maintain or improve the economic, health, cultural, and social standards of the areas that the mining project will impact. Included are not only the physical infrastructure (buildings, equipment) but also the resources (personnel, supplies) needed to maintain and operate these institutions. The quality of the social infrastructure will in turn impact the effectiveness of the mining project. The mining company should assess with the local and central government the need to improve the quality of the social infrastructure and to determine who is going to pay for these improvements. The mining company may have to supply such services directly and at its own cost.

One might expect that the increased economic activity which mining will bring to the community, and the taxes which government will collect, would supply sufficient funds to upgrade the social infrastructure. However, there is a significant time lag between project construction, when the need for improved social infrastructure begins to be felt, and positive, taxable cash flow. Furthermore, tax benefits are mostly accrued by the central government, which has to satisfy multiple constituencies competing for funds, of which the mine's district is only one.

Most mining companies recognize and publicly state their corporate social responsibility toward environment and stakeholders. While there is a direct financial cost in fulfilling this responsibility, the benefits are in the form of obtaining and maintaining the social license to operate by reducing the risk inherent to opposition to project construction or operations. Items that are part of the needed social infrastructure include

- Childcare, schools, and education facilities, whose improvement can significantly enhance the public image of the mining company while having visible short- and long-term benefits to the local population;

- Healthcare facilities and emergency services, accessible to both the mine and the community;
- Law enforcement and security services to control the effect of an increased, transient, and often disproportionally male population;
- Food supply as needed to satisfy the rapid increase of population and change in demand patterns;
- Road quality and transportation services other than those directly required by the mine, which may need to be improved to accommodate increased traffic;
- Sewage system and treatment facilities;
- Potable water distribution network and wastewater treatment facilities, which are likely to have the most significant impact on the health of the community as a whole, including employees living in the community or using local facilities;
- Telecommunications network accessible outside the mine area (Internet, phone, television);
- Recreation facilities;
- Hotels and housing infrastructure as needed for temporary and permanent staffing; and
- Training facilities needed to make it possible to recruit locally the human resources needed to build, maintain, and operate the mine, as well as the new social infrastructure.

RISK MANAGEMENT AND INFRASTRUCTURE

The role of those responsible for the design and evaluation of the necessary infrastructure is to supply estimates concerning technical requirements and financial, social, and environmental costs. These estimates take into consideration information supplied by others, as well as company experience, knowledge, and judgment. Whether the infrastructure requirements and costs are estimated by the mining company or, as is often the case, by an independent EPC company, the risk associated with these estimates must be understood

ASSESSMENT OF INFRASTRUCTURE REQUIREMENTS

The need for effective communication between geologist, mining engineer, process engineer, and infrastructure engineer were discussed in the Chapter 13 sidebars. An example follows of a situation where lack of expert due diligence resulted in significant cost underestimation and project failure.

- After cursory review of information supplied by the vendor, a mining company acquired a mineral deposit located in a remote part of Africa. Following acquisition, the buyer's engineers visited the site and completed a scoping study aimed at determining development options. The study showed that the deposit properties were sufficiently well defined and that the mining and processing options suggested by the vendor were technically feasible. However, a review of existing and needed infrastructure showed a very different picture:
  - Existing access infrastructure was minimal. Roads were designed for local commerce, which consisted mostly of animal-drawn carts, a few cars, and light trucks.
  - Power supply was nonexistent. The closest high-voltage electric line was 100 km away and supply was unreliable with frequent interruptions. The processing plant alone would consume all the power available.
  - The closest harbor was 250 km away and a railroad line would have to be built from the mine to the coast. The line would cross two countries, which would require undefined permits from two governments. Government expectations were that the mining company would build the line at its own cost, but access by others should be allowed.
  - The existing harbor facilities were insufficient and would have to be significantly upgraded.
  - Geographic and geologic information concerning the lands on which the railroad line would be built were of insufficient quality to estimate construction costs beyond the scoping study.
  - Villages, agricultural land, rivers, and streams would be affected by infrastructure construction with undefined social and environmental costs.
- After completing the review, the mining company determined that the financial, political, and social risk was too high to justify project development. The cost of acquisition and feasibility study was written off and the project was abandoned.

and accepted by the mining company. The mining company must start by making sure that the estimators have necessary competence. Experience must be relevant to the project's geologic, geographic, political, environmental, and social surroundings. Uncertainty concerning infrastructure requirements must be quantified, including the probability that additional infrastructure might be required as conditions change over time. Costs must be estimated accordingly and taken into account when estimating the overall project risk.

Risk factors associated with infrastructure design and construction were discussed earlier. A particular risk relates to external infrastructure and infrastructure built by contractors. The mining company must take responsibility for all aspects of a project, even if this responsibility is legally and contractually transferred to contractors or an EPC company. Entities outside the company, including local communities, governments, nongovernmental organizations, and other groups that monitor national and international mining projects, do not differentiate between the project owner and those working for the project owner. The social and environmental impact of any work related to the project is perceived by all external entities as being the responsibility of the owner. Failure to understand the difference between legal and perceived responsibilities can have dire consequences on public acceptance of the project.

CHAPTER 17

MANAGEMENT

It is decreed by Divine Providence that those who know what they ought to do and then take care to do it properly, for the most part meet with good fortune in all they undertake; on the other hand, misfortune overtakes the indolent and those who are careless in their work.

Georgius Agricola, *De Re Metallica*, 1556

The success of a mining company is a function of its ability to

- Discover or acquire mineral deposits that present reasonable prospects for eventual economic extraction;
- Reliably evaluate the economic feasibility of developing such deposits;
- Develop these deposits as planned, on time, within scope, and within budget;
- Operate the mines in a financially and environmentally sustainable fashion;
- Successfully bring a project to closure; and
- Complete these tasks while respecting environmental, social, political, legal, regulatory, and ethical obligations.

Success or failure in achieving these objectives is dependent on management's ability to develop and maintain a work environment in which highly competent persons work toward a well-defined common objective, open and honest communications are expected and rewarded, and there is no tolerance for unethical behavior.

CORPORATE MANAGEMENT

The chief executive officer (CEO) is the public image of the company and plays a critical role in maintaining the company's social license to operate and long-term sustainable viability. The CEO sets the company's strategy, vision, and values; ensures required corporate governance; builds the company culture; and hires a management team that endorses and propagates this culture. The CEO allocates human and financial resources from the company, and ensures compliance with all legal and regulatory requirements.

The CEO is responsible for final decisions concerning development of projects requiring major investments. To make these decisions, the CEO relies on information and recommendations made by his or her executive team, the project managers, and the department heads responsible for assessing the technical, financial, environmental, and social aspects of the project. There are many levels at which decisions are made by the team, from the junior geologist who supervises the daily operations of drillers during exploration, to the project manager who must take responsibility for all decisions and recommendations made by his team. At each level, data are collected and interpreted, uncertainty is assessed objectively or subjectively, and recommendations are made that form the basis for decision-making at the next level.

The ultimate decision made by the CEO is dependent on the confidence that he or she can give to the information received, the uncertainty associated with this information, and the risk taken by choosing one decision over another. The CEO may have to present his or her recommendations to the board of directors for final approval. Provided regular communications are maintained with the board, no change in direction is expected when final recommendations are made.

MISSION AND VALUES

An important role of the CEO is to ensure that the company culture and actions reflect his or her vision, and that the image that is projected both within and outside the company reflects reality. The mission statement of all public companies, whether they sell beauty products, consulting services, automobiles, gold bullion, or copper concentrate, starts with recognition of their financial responsibility to the shareholders. While reassuring to the

MANAGERIAL DECISIONS AND RISK MITIGATION

Contracts between the mining company and suppliers of goods and services might be considered as effective ways to mitigate risk. However, it may be unwise to assume that contracts fully transfer risks to others. When agreeing to supply goods and services, the supplier, whether a private company or a government entity, is taking risks of its own. It is the mining company's responsibility to understand these risks and to evaluate the extent to which they are likely to affect the supplier's ability to fulfill its contractual obligations.

The effectiveness of communication should not be measured solely by its frequency. Emails are generally considered as an easy way to ensure effective communication. This is not necessarily the case. Indiscriminately sending emails to everyone working on a project is not an efficient way to communicate. Such emails may not be read by any of the recipients. Conversely, they might be read by all, resulting in considerable loss of time. If there are 20 recipients and each one takes one minute to determine whether an email concerns him or her, 20 minutes are wasted per email. If 30 emails are received daily, 10 hours are wasted per day. Assuming 250 working days per year and a cost per employee of $200 an hour, the loss is $500,000 per year. In the worst scenario, the one person who needs to act according to the content of the email could be the one who does not read it.

shareholders, such a statement does not help discriminate between companies. Most public companies also make statements to the effect that they are committed to sustainability, respect of the environment, and social responsibility. Such statements may be considered as truisms because they reflect not only company values but also company legal requirements. It still remains that in most situations, management and employees do live and act in accordance with the stated mission and values of the company.

It is not sufficient for the company to publish and disseminate statements concerning core values. Management must show that it truly believes in these values, and it must act accordingly. Behaviors that are consistent with these values should be rewarded while those that are not should be corrected. Leading by example is key. Many examples can be found of situations where personal interest and short-term gratification were put ahead of the company's interests and the interests of shareholders, and decisions were made in violation of company values.

Behavior and actions that are not consistent with the public image that the company wants to project can have serious consequences, not only for the individual involved—who presumably is aware of and accepts the risk taken—but also, and more importantly, for the company as a whole. Corporate governance structures must be put in place to ensure that such behavior does not take place, and if it occurs, that it does not remain undetected. Determining the most effective structures is particularly challenging, because misbehavior can occur at all levels while effective management requires an open environment based on trust.

CORPORATE GOVERNANCE

There are many definitions of what *corporate governance* means. In a narrow sense, corporate governance relates to the allocation of powers between shareholders and the board of directors, and how investors ensure that the company's decisions reflect their interests and expectations. Corporate governance identifies the context within which the company's objectives are set and how these objectives are to be reached, taking into account governmental, legal, regulatory, environmental, and social obligations. As it relates to risk tolerance, corporate governance must specify what level is acceptable. Risk management forms an integral part of corporate governance.

A broader definition considers corporate governance as the mechanisms, rules, practices, values, and processes by which the company is directed and controlled, taking into account the interests of all stakeholders, including not only the shareholders but also financial institutions, employees, customers, governmental and nongovernmental organizations, and the social and biological environment. These definitions of corporate governance are not exclusive but complement each other. The shareholders' interests, which might be primarily of a financial nature, cannot be satisfied without due consideration being given to the potentially conflicting interests of all other stakeholders.

Corporate governance must define the context within which mineral resources are evaluated, including the following:

- Role of the board of directors, chief executive officer, and senior management team

- Role of the project manager
- Role of the project technical team, including competent persons responsible for geology, mining, processing, and infrastructure
- Role of other company departments that support the project team, such as legal, social, environmental, financial, governmental, and risk management

SUSTAINABLE DEVELOPMENT PRINCIPLES

A basic requirement for the successful development and operation of a project is that the mining company acts in a sustainable, ethical, and responsible fashion. The company must demonstrate that it understands its responsibilities toward the environment, both ecological and social, in which it operates. A company may have acquired all necessary governmental permits and authorizations to go ahead with a project, but this alone is not sufficient to obtain support from local communities, indigenous people, nongovernmental agencies, and other stakeholders. Management must clearly define its corporate social responsibility and how this responsibility should be taken into account in the design and evaluation of new projects. The social license to operate (SLO) is not likely to be obtained without demonstrated commitment to the environment in which it intends to operate.

The cost of corporate social responsibility can be significant and must be included in the evaluation of mining projects. Failure to obtain the SLO can stop development of a new project even if its operation was expected to be highly profitable to the shareholders. Failure to maintain this social license can also stop operations at any time during the life of a project. Many mining companies have suffered significant losses resulting from failure to obtain or maintain the necessary SLO from local communities. The financial cost of these losses is not limited to the specific project under consideration. The speed and ease with which news and rumors are exchanged worldwide are such that loss of SLO in one area can affect company operations worldwide.

The International Council on Mining and Metals (ICMM) is a membership organization led by the CEOs of many of the world's largest mining and metals companies and associations. ICMM is committed to driving social, economic, and environmental progress. Its vision is one of leading companies working together and with others to strengthen the contribution of mining, minerals, and metals to sustainable development. In 2003, ICMM committed member companies to implement and measure their performance against 10 sustainable development principles, which were updated in 2015 (ICMM 2015):

1. Apply ethical business practices and sound systems of corporate governance and transparency to support sustainable development.
2. Integrate sustainable development in corporate strategy and decision-making processes.
3. Respect human rights and the interests, cultures, customs and values of employees and communities affected by our activities.
4. Implement effective risk-management strategies and systems based on sound science and which account for stakeholder perceptions of risks.
5. Pursue continual improvement in health and safety performance with the ultimate goal of zero harm.
6. Pursue continual improvement in environmental performance issues, such as water stewardship, energy use, and climate change.
7. Contribute to the conservation of biodiversity and integrated approaches to land-use planning.
8. Facilitate and support the knowledge base and systems for responsible design, use, re-use, recycling and disposal of products containing metals and minerals.
9. Pursue continual improvement in social performance and contribute to the social, economic, and institutional development of host countries and communities.
10. Proactively engage key stakeholders on sustainable development challenges and opportunities in an open and transparent manner. Effectively report and independently verify progress and performance.

Mining companies are expected to use these sustainable development principles to define corporate social responsibilities. CRIRSCO, the Committee for Mineral Reserves International Reporting Standards, whose role was discussed in Chapter 4 is a strategic partner of ICMM.

PROJECT MANAGEMENT

Successful evaluation of the economic potential of a mineral deposit requires nomination of a project manager whose responsibilities are clearly defined. Key project management responsibilities include the following:

- Be knowledgeable of rules, regulations, guidelines, and best practices that apply to the evaluation of mineral deposits.
- Understand the strategic objectives of the company and how the project is expected to contribute to reaching those objectives.
- Have an understanding of all aspects of the mining cycles, from exploration to closure, from years of high prices to years of depression, from technical factors to human factors, from financial to governmental, from legal to political, and so forth.
- Have a sound appreciation of areas of personal weakness, and the ability to compensate by forming a management team whose members complement each other.
- Implement systems and procedures that create independent channels of communication to verify that the normal channels are open, honest, and accurate.
- Define specific tasks to be completed to reach the project objectives and estimate resources needed to complete these tasks, including staff, budget, and schedule. The first such estimates are preliminary, to be presented to senior management to ensure mutual agreement on scope and objectives.
- Review project objectives, scope, deliverables, schedule, and budget with corporate management and get approval to go ahead with project evaluation.

HOW KPIS AND BONUSES CAN INCREASE RISK

Key performance indicators (KPIs) and related bonuses play an important role in improving performance. However, if poorly designed, they can produce unexpected and undesirable results. An example follows.

In an operating mine, operators were informed at the beginning of the year that their bonus would be a function of total tonnage mined, cost per metric ton mined, and tonnage and grade of material sent to process. The underlying but unstated assumption was that this would be achieved within the bounds of the specified mine plan. The operators realized that they could optimize their bonus by taking into account the following strategies:

- Tonnages mined could be increased and mining costs could be decreased by accelerating pre-stripping and minimizing tonnage from deeper parts of the pit.
- The high-grade material was at the bottom of the pit. High-grading would result in satisfying the tonnage required by the processing plant and exceeding the minimum grade requirement.
- A tailings dam expansion was under construction using mine waste. The cost of trucking the waste from the rim of the pit to the waste dam was treated as a dam construction cost rather than a mining cost. Accelerating pre-stripping and sending this material to the tailings dam significantly reduced the costs allocated to the mine.

Sidebar continues next page

- Build the team that will complete the project under the project manager's leadership. In collaboration with this team, define individual roles, deliverables, schedules, and budget.
- Define processes for internal and external communication, quality control, risk assessment, project change avoidance, and change management if required.
- Present revised deliverables, schedule, and budget to corporate management for final approval.
- Manage the project from authorization to proceed to completion and closure.

HOW KPIS AND BONUSES CAN INCREASE RISK

These actions resulted in the following outcomes:

- Total tonnage mined was higher than planned and costs were lower, resulting in increased bonuses that were significantly higher than budgeted.
- Material processed was higher grade than planned, resulting in higher revenues from sales and higher bonuses.
- Mining activities that were planned at the bottom of the pit to expose ore for future processing were delayed. The ore that was exposed was high-graded. Present costs were reduced but future costs would be higher. The next-year production schedule would have to be changed. Future mill feed would be a lower grade than planned.
- Accelerating near-surface pre-stripping and sending the waste to the tailings storage facilities resulted in the dam being built faster than necessary with a higher capitalized cost of construction.

The actions taken by the mine operators were consistent with their objective, which was to maximize their bonus. When compared with the mine plan, the current-year mining costs were reduced and revenues from sales were increased. However, the following year's mining costs would be higher, and revenues from sales would be lower. These outcomes benefited the individuals but not necessarily the company as a whole.

- Coordinate activities of the different disciplines, ensuring effective communication and utilization of available resources to reach goals and objectives in a timely fashion. Weekly or quarterly meetings with discipline heads and relevant experts will serve to make sure that the team objectives remain in line with expectations. The frequency of the meetings, and their duration, should vary depending on the project status. One of the challenges faced by the project manager is to ensure that only those who need to attend the meetings are present. Meetings with dozens of attendees are not likely to be productive.
- Particular attention should be given to budget, schedule, quality or reliability of results obtained, and decisions made. The ability to predict and resolve issues is one of the key requirements for effective

project management, whether these issues are of a technical, financial, budgetary, scheduling, resource allocation, staffing, or other nature.

- Maintain periodic communications with senior management, including progress updates and budget and schedule statuses.
- Schedule or budget changes should be avoided. Management should be kept informed if the necessity for such changes appears likely, and authorization should be obtained before changes are made.
- Once the study is completed, present results to senior management, including recommendations for a path forward and the risk associated with these recommendations. If the study considered more than one option, such as during the pre-feasibility study of a project, each one should be summarized and the recommended option should be justified.

Project study documentation should be a company requirement. The documentation process should take place from project initiation to closure. It should be considered a key control as well as a necessary part of the company's learning process. Documentation should describe not only decisions and recommendations but, as importantly, how these decisions were made. Only if a study was properly documented will it be possible to compare future performance with original expectations. There are minimum legal requirements for documentation, such as those specified by stock exchange regulators (e.g., NI 43-101 in Canada).

The project manager is not expected to be an expert in all the different disciplines that contribute to the evaluation of a mineral deposit. However, he or she should have sufficient knowledge of those disciplines to be able to assess the competence of the project team, to evaluate the quality of the work being done and the risks being taken, to ensure the reliability of estimates developed by the team, and to call for additional expertise where gaps are detected.

PROJECT STAGES

Examples of resource evaluation projects are given below that may call for different qualifications on the part of the project manager:

MANAGERIAL DECISIONS WITH COSTLY OUTCOMES

Situations where management can motivate behavior detrimental to the company's interests were presented earlier in a Chapter 12 sidebar. Other examples follow.

Case 1

The chief executive officer (CEO) of a mining company projected an image of honesty, ethical behavior, and social and environmental responsibility. Meanwhile, his chief operating officer (COO) was consistently demanding that the mine operators focus their entire attention on actions aimed at improving share price and maximizing bonuses. Consequently, risky decisions were made on-site that did not respect the company's values and sustainable development principles. The image presented by the COO to the CEO and the board of directors was that of following all U.S. standards of safety, ethical, environmental, and social behavior. This example shows the challenges that are faced by management at all levels. Honest communication is expected but cannot always be taken for granted. Multiple communication channels are needed to ensure the quality of information received.

Case 2

A member of the board of directors overheard a conversation between the COO and the general manager (GM) of an operating mine. The GM was concerned that the presentation he was to make to the board showed increased profitability without mentioning a concurrent decrease in operational safety. The COO's answer was that he was aware of the risk and that it did not need to be mentioned to the board. The board member who overheard the conversation failed to interrupt it or to openly discuss the situation with the board and its ethics committee. This example illustrates the need for a board of directors that is composed of strong personalities and has the power to get information outside the normal channels of communication.

Case 3

Senior management was presented preliminary estimates of end-of-year mineral reserves and financial results. These estimates were declared to be unacceptable because they differed negatively from market expectations. The managers in charge of reserves and financial reporting were told to go back to the drawing board and that "better" results were expected to be presented at the next meeting. Besides being unethical, such a request could have resulted in risky and potentially illegal decisions being made by the respective managers. The correct approach should have been to determine how to best assist the managers in a systematic review of the estimates, with the understanding that the final results would be accepted as they were.

- **Evaluation for acquisition.** Evaluation of the economic potential of a mineral property that is considered for acquisition is usually subject to considerable time constraints. Only a few days or weeks might be available to review available information and recommend whether to purchase the property and at which price. Information is likely to be limited, especially in hostile takeovers. The quality of the available information should be questioned, as it may be biased. The interest of the seller is to emphasize the positive aspects of the project while deemphasizing the negatives. The buyer should not overestimate his or her own ability to add value to the project above that claimed by the seller. The ideal project team should be led by a manager who has proved prior experience in business acquisition and is supported by a small team of senior technical experts. The project manager needs to understand why the acquisition is important, and to make recommendations to senior management that fairly take into account risks and opportunities.

- **Early-stage evaluation and scoping study.** A scoping study is an order-of-magnitude technical and economic study of the potential viability of mineral resources. Scoping studies are performed when a mineralization has been discovered but only limited geologic information is available. The objective is to decide on a path forward: Should exploration continue, and if so, what resources should be allocated to it? A scoping study is likely to be completed in a few days or weeks by a small team of company experts. The decisions depend on the likelihood that the project will be of interest to the company if additional information is made available. Uncertainty originates primarily from insufficient geologic information. Only order-of-magnitude estimates of capital and operating costs can be developed.

  The project manager and his or her team are expected to have a strong technical background as well as experience concerning capital and operating costs for similar projects. Recommendations made on the basis of a scoping study can be risky. One might decide to continue exploration of a deposit that proves uneconomic, or one might prematurely abandon a project of significant but unrecognized economic potential. Successful exploration geologists tend to be very optimistic individuals. Mine geologists as well as mining and metallurgical engineers are less so and can be overly conservative. Controlled optimism is a quality that the project engineer must have to balance these opposite views of a project's potential and thereby reach sound decisions. A scoping study must be completed before an inferred resource is publicly declared.

- **Advanced stage and pre-feasibility study.** A pre-feasibility study is a comprehensive study of a range of options for the technical and economic viability of a mineral project. At the end of a pre-feasibility study, if the results show that the deposit can be developed economically, a single option is chosen to form the basis on which a detailed feasibility study will be completed. The available information must be detailed enough to make comparison of various technical and economic options meaningful. The geologic understanding of the deposit must be such that all or part of the deposit can be classified as a measured or indicated mineral resource and used to estimate a mineral reserve.

  Deposit models must be available to develop mine plans and production schedules. Although the same drill-hole information is used for each model, the models may differ depending on the mining and processing methods under consideration, and all models must be compatible with each other. Geometallurgical information must be sufficient to define feasible processing methods and estimate corresponding recoveries, throughput, and capital and operating costs. The political, social, environmental, legal, and regulatory environment must be understood and not present any fatal flaws. A pre-feasibility study is likely to require input from dozens of individuals and may last for months.

  The role of the project manager is no longer likely to include a significant technical component. A project engineer might be nominated as a liaison between the project manager and the technical disciplines involved in the project. The project manager needs to give particular attention to schedule and budget, scopes and deliverables, team building and motivation, resource management, communications between disciplines, risk assessment, and communications with senior management.

- **Final stage and feasibility study.** A feasibility study is a comprehensive technical and economic study of the selected development option for a mineral project. The key word is comprehensive. All aspects of the project must be thoroughly analyzed. Risk assessment, control, and mitigation are an integral part of such a study. The technical and financial part of the study must reach the level of detail needed to estimate capital and operating costs with a high level of confidence. Completing the feasibility study of a large mining project is likely to cost millions of dollars and last many months or even years. Hundreds of people may be involved. Staffing is likely to include employees of the mining company as well as external consultants and one or more engineering, procurement, and construction (EPC) companies.

  The project manager's main responsibilities include organization, staffing, scope of work, definition of clear and attainable objectives, selection and performance management of consulting and EPC firms, budget, schedule, quality control, risk assessment, breaking silos, coordinating work between teams, and communication with vendors. Senior management must be continuously and effectively informed of progress made and critical issues discovered as they occur. A key role of the project manager is to ensure effective communication between mining company representatives and service suppliers. The project manager must create an environment in which trust and open communication are expected. The project team must feel confident that issues can be discussed openly and without retaliation. The project manager should have a proven record of success in developing mineral projects of the size and type under consideration. Relevant experience is needed not only with project management but also with mining operations to be able to evaluate, understand, and trust decisions and recommendations made by key members of the project team.

MANAGERIAL DECISIONS AND WHISTLEBLOWER

In an operating mine, all reports of incidents and accidents were posted in the lunchroom to make employees aware of safety hazards and the need to be safety conscious. These reports showed that most accidents, some of them serious, were caused by rock falls. The chief mining engineer expressed her concern to the mine manager and recommended that the mine plan be modified to reduce risk. This would require flatter pit slopes and additional waste stripping and would result in an increase in mining costs and initial decrease in production. The mine manager's response was that productivity could not be compromised because the project economics were deteriorating quickly as a result of falling metal prices.

The mining engineer disagreed with placing profitability ahead of safety but received no support from local management and was told to follow orders. Being aware that taking no action could have serious ethical and legal consequences, she decided to contact the head office management directly. After reviewing the situation, the chief operating officer decided to hire an international consulting company to review safety conditions at the mine. The opinion of the consulting company was that, if the mine was inspected by government agencies, it would be shut down immediately and the company would be severely penalized. The mine manager left the company "to pursue other opportunities," and the mining engineer was transferred to another mine.

This example shows the importance of having effective and enforceable whistle-blower policies. It also shows that routine inspection of the mine sites, by independent internal or external experts reporting directly to senior management and the board of directors, should be part of the governance process. Such inspections would help ensure that management is always getting reliable information, would significantly reduce any temptation to make unsound decisions for short-term benefits, and would ultimately eliminate the somewhat unreliable dependence on whistle-blowers.

MANAGEMENT AND RISK CREATION

A key role of management is to ensure that the project risk is properly assessed and that appropriate risk mitigation plans are in place. To become effective leaders, managers must have high confidence in their proven ability to make sound decisions under uncertainty while keeping a healthy dose of realistic skepticism. CEOs and senior managers are expected to demonstrate their ability to act decisively with limited and sometimes unreliable information. A CEO's success is dependent on the trust that stakeholders believe they can place in his or her management. A manager's likelihood of being promoted, or even to keep his or her job, is dependent not only on proven success but also on the ability to project an image of reliability. It may therefore appear odd that in many instances management practices can increase risk. The following examples illustrate situations where management decisions result in risk creation:

- Failure to assign a project manager as a single point of responsibility for the entire project
- Poor understanding of the project manager's role, resulting in assignment of under-qualified managers.
- Poorly defined or conflicting study objectives, resulting in lack of direction and misallocation of company resources
- Specifying objectives and setting performance targets that directly or indirectly define what the outcome of the study should be, thus compromising the integrity of the results
- Unfounded management expectations that, if communicated to the project team, result in biased studies
- Lack of appreciation of the relationship between project risk and resources allocated to the project, resulting in excessive or insufficient resources being assigned
- Unrealistic planning, budget, and schedule, the consequences including completion delays, cost overruns, or excessive risk-taking
- Project team lacking specific expertise necessary to analyze the project, and reluctance in using external resources when qualified internal resources are not available
- Mining company lacking internal resources, qualifications, and project controls to properly manage external consultants and EPC companies
- Ineffective project management controls and cost accounting
- Project manager not recognizing the formation of silos and lack of communication between disciplines

- Failure of project manager to supply senior management with information needed to make timely decisions concerning necessary changes in scope, schedule, or budget
- Poorly designed motivators, resulting in undesirable behavior, such as bonus plans that reward management for proving that a project should be developed
- Bonus plans and rewards that motivate the project team to demonstrate that the project is economically feasible instead of fairly analyzing its value
- Motivating unethical behavior, such as encouraging overestimation of quantity and quality of a mineral resource to increase its apparent value to a potential buyer
- Motivating EPC contractors to show a project as being economic if such an outcome is necessary to satisfy publicly stated company objectives
- Lack of processes to prevent unethical or illegal behavior or to bring such behavior to the attention of management, such as whistleblowers being ignored, sidetracked, or otherwise silenced
- Company values and ethics being defined and emphasized in annual reports and press releases but ignored in day-to-day decisions and actions

Many other situations can be found where management practices at various levels are detrimental to the sound evaluation of projects and to the overall long-term success of the company.

MANAGEMENT AND RISK MITIGATION

Risk management is the systematic process of understanding, evaluating, and addressing project risks to maximize the chances that specific objectives will be reached, including those of a financial, environmental, and social nature. Risk management should also exploit the opportunities that uncertainty may bring as new possibilities present themselves during the project life. Effective risk management requires an informed understanding of relevant risks, an assessment of their relative significance, and a rigorous approach to monitoring and controlling them.

Risk evaluation and risk mitigation can be a difficult, long, and costly process. Depending on situations, the approach to risk that a company might take includes risk ignorance (one single value is developed), risk assessment (risk is quantified but the decision to go ahead does not take it into account), or risk management (a detailed study is completed and risk mitigation plans are developed).

- **Risk ignorance:** If risk is ignored, only one estimate is developed, the assumption being that it is accurate and that estimation errors can be ignored. This approach may be reasonable for small projects whose failure would not be of material significance. In public releases of project information, single estimates are often presented. Mineral reserves are published as a single number. Public statements are made about risk as required by regulations and guidelines, but these statements tend to be of a generic nature and do not give the reader a clear indication of how the risk factors that are listed are likely to influence the specific project under consideration.

- **Risk assessment:** Risk assessment is a passive approach to risk. It consists of determining the risk related to choosing a given option but assumes no flexibility to the project as conditions change over time. This approach is used to determine whether the risk associated with a project as designed is acceptable; if it is not, the project might be redesigned to reduce risk. Risk assessment is the method most commonly used when presenting projects to management for approval. If done properly, this approach will overestimate risk because it ignores the possibility of change management.

- **Risk management:** Risk management requires an active approach to risk mitigation, including implementation of effective crisis management and business continuity plans. The preferred project should be one that can adapt positively to unforeseen changes to the operating environment as and when required. The project risk is estimated taking into account when events might occur, what the consequences would be if they occur (risk assessment), and what actions could be taken to mitigate the consequences (risk management).

The following steps can be taken to determine the preferred option:

1. Identify the risk variables or events most likely to influence the economic feasibility of the project.
2. Determine the values that the risk variables might take and the probability of occurrence of these events.
3. Determine the consequences that may result if such events do occur. The project utility can be determined by combining probability and consequences. This project utility assumes that no mitigating action can be taken.
4. Identify ways to reduce the consequences. Evaluate the cost of taking appropriate mitigating actions.
5. Reestimate the utility of the project taking into account (a) the probability that events will occur, (b) the cost of mitigating the consequences, and (c) the remaining impact that the events would have after mitigation.
6. To mitigate risk most effectively, develop a crisis management plan to react quickly in the event that a crisis does occur.

CHAPTER 18

CONCLUSIONS

Risk is part of life. Avoiding all risk would result in no achievement, no progress, and no reward. There is nothing exceptional about how mining risk should be assessed and decisions should be made. The methods to be used and processes to put in place are not different from those that apply to other industries. A systematic process should be developed that ensures proper estimation of risk factors—determination of which events, favorable or unfavorable, may occur; the probability that they will occur; the consequences that such events may have if they occur; and the mitigating actions that should be taken to minimize the effect of these events if negative or maximize their effect if positive.

To be effective, mitigating plans should be prepared in advance, even if events that might occur during the mine life are unknown. There is uncertainty concerning which event might occur, when it will occur, and what will be its consequences. But one should not wait for an event to occur (or to have a high probability of occurrence) to start developing an action plan. Processes should be put in place so that everybody knows what they should do if events occur to ensure that mitigating plans are effectively and promptly executed. Any delay can have critical consequences.

Management must ensure that all those responsible for project valuation are fully aware of the need to develop reliable estimates that are not only precise, unbiased positively or negatively, but also accurate, with acceptable and manageable associated uncertainty. One of the highest risks results from excessive confidence in the competence of management and the project team. Quality control and quality assessment of project evaluation should be based on internal and external independent reviews of critical aspects of the project. When choosing the reviewers, one should be aware that they, like the original project team, have their own strengths and weaknesses, and that they may also not be fully aware of their weaknesses, or be willing to recognize them. The reviewers must add to the project team, not duplicate its capabilities.

Risk management is not easy. The information needed to effectively evaluate the risk taken when making major investment decisions is not easily obtained. It takes time and costs money. Considerable experience in risk assessment is needed, not only from those who develop corporate models but also from all who supply information needed to make these models meaningful. But the rewards are worth the effort. As discussed throughout this book, ignoring or underestimating risk has resulted in billions of dollars being wasted by the mining industry. Unless we learn from the past, we are bound to repeat the same mistakes. At the time this book was completed, in mid-2016, gold prices were once again showing rapid increases, comparable to those observed in 2010 and 2011. Will the lessons learned in 2013, when gold prices collapsed, be remembered, or have they already been forgotten? It is hoped that this book will contribute somewhat to ensuring that rational decisions are made, even when market pressures create an environment in which the temptation for immediate but highly risky actions can be very hard to resist.

APPLICATION OF MONTE CARLO SIMULATION TO A COPPER-GOLD DEPOSIT

In this example, a copper-gold deposit is considered for development, which is located next to an operating mine. Only a few drill holes have been completed to date. The question under consideration is whether additional drilling is warranted. There is insufficient information to develop a three-dimensional model. The size of the deposit is considered reasonably well defined, but the grade is highly uncertain. To evaluate the risk associated with the project, the first step consists of estimating the probability distribution of the copper and gold grades.

Using the available drill-hole data and additional information obtained from the nearby operating mine, the probability distribution of the deposit grade is as shown in Table A.1, which reveals the following:

- x_1 represents the copper grade and x_2 the gold grade.
- The bottom two rows show the range of possible copper grades, from $x_1 = 0.1\%$ Cu to $x_1 = 0.5\%$ Cu.
- Above the copper grades are the likelihoods $p(x_1)$ that theses grades will be realized. There is only 5% probability that the copper grade will be as low as 0.1% Cu. The most likely grade is 0.3% Cu with a probability of 46%. Note that the sum of these probabilities must add up to 1.0.

- The first two columns on the left-hand side show the range of possible gold grades, from $x_2 = 0.01$ g/t Au to $x_2 = 0.16$ g/t Au.
- Next to the gold grades are the likelihoods $p(x_2)$ that theses grades will be realized. There is only 8% probability that the gold grade will be as low as 0.01 g/t Au. The most likely gold grade is 0.04 g/t Au with a probability of 33%. Note that the sum of these probabilities must add up to 1.0.
- Gold grades and copper grades are not independent. If the copper grade is high, the gold grade is also generally expected to be high. The correlation between grades must be modeled before one can simulate a realization of both copper and gold grades. The probability $p(x_1,x_2)$ that the copper grade will be x_1 and the gold grade will be x_2 is shown under the $p(x_1,x_2)$ header. The table shows that there is only 3% probability that the copper grade will be 0.10% Cu and the gold grade 0.01 g/t Au. The most likely grade combination is $x_1 = 0.3\%$ Cu and $x_2 = 0.04$ g/t Au, which has 18% probability of being realized.
- The sum of all $p(x_1,x_2)$ probabilities must add up to one.
- If we add these probabilities by column (constant x_1), they must add up to the corresponding probability $p(x_1)$. For example, the values of $p(x_1,x_2)$ above the copper grade $x_1 = 0.3\%$ Cu add up to 46%.
- If we add these probabilities by row (constant x_2), they must add up to the corresponding probability $p(x_2)$. For example, the values of $p(x_1,x_2)$ to the right of the gold grade $x_2 = 0.04$ g/t Au add up to 33%.

To realize a set of copper and gold grades, one must first simulate a value of x_1 using $p(x_1)$ and then simulate a value of x_2 using the probability distribution $p(x_2|x_1)$ of x_2 given x_1. This probability is calculated using the following formula:

$$p(x_2|x_1) = p(x_1,x_2)/p(x_1)$$

The results are given in Table A.2. In this table, the columns add up to 100%. If the copper grade is 0.3% Cu, the probability that the gold grade is 0.04 g/t Au is 39%, calculated as follows:

$$p(x_1,x_2) = p(0.3\%\ \text{Cu}, 0.04\ \text{g/t Au}) = 18\%$$

$$p(x_1) = p(0.3\%\ \text{Cu}) = 46\%$$

$$p(x_2|x_1) = 18\% \,/\, 46\% = 39\%$$

The next steps consist of the following:

1. Using $p(x_1)$ from Table A.1, calculate and plot the cumulative probability distributions (cpd) of x_1, $cpd(x_1)$. The results are given in Table A.3 and plotted on Figure A.1
2. Generate a random number between 0 and 1 from a uniform distribution. For this example, it is assumed that the generated random number is $r_1 = 69\%$.

TABLE A.1 Probability distribution of copper and gold grades

x_2		$p(x_2)$	$p(x_1,x_2)$					
Gold Grade (g/t)	0.16	7%	0%	0%	4%	3%	0%	$p(x_1,x_2)$
	0.08	24%	0%	2%	13%	8%	1%	
	0.04	33%	0%	3%	18%	10%	2%	
	0.02	28%	2%	7%	9%	9%	1%	
	0.01	8%	3%	2%	2%	1%	0%	
		100%	5%	14%	46%	31%	4%	$p(x_1)$
			0.1	0.2	0.3	0.4	0.5	x_1
			Copper Grade (% Cu)					

TABLE A.2 Probability distribution gold grades given the copper grades

x_2		$p(x_2\|x_1) = p(x_1,x_2)/p(x_1)$					
Gold Grade (g/t)	0.16	0%	0%	9%	10%	0%	$p(x_2\|x_1)$
	0.08	0%	14%	28%	26%	25%	
	0.04	0%	21%	39%	32%	50%	
	0.02	40%	50%	20%	29%	25%	
	0.01	60%	14%	4%	3%	0%	
		100%	100%	100%	100%	100%	
		0.1	0.2	0.3	0.4	0.5	x_1
		Copper Grade (% Cu)					

TABLE A.3 Cumulative probability distribution of x_1

Copper Grade	x_1 (%Cu)	0.10	0.20	0.30	0.40	0.50
Probability of occurrence	$p(x_1)$	5%	14%	46%	31%	4%
Cumulative distribution	$cpd(x_1)$	5%	19%	65%	96%	100%

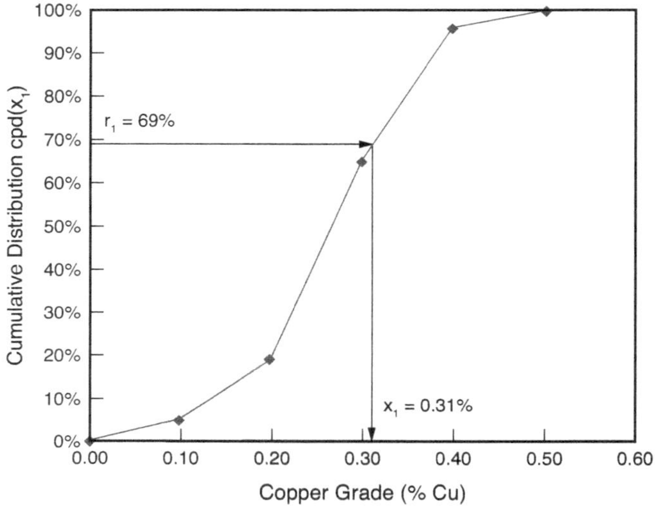

FIGURE A.1 Cumulative probability distribution of x_1

3. Determine the value of x_1 for which $cpd(x_1) = r_1$. As shown in Figure A.1, this value is $x_1 = 0.31\%$ Cu.

4. Using $p(x_2|x_1 = 0.31\%\ \text{Cu})$, calculate and plot the cumulative distribution of x_2, $cpd(x_2|x_1 = 0.31\%\ \text{Cu})$. As there is no significant difference between 0.30% Cu and 0.31% Cu, the readily available probabilities corresponding to 0.30% Cu will be used. The results are given in Table A.4 and plotted on Figure A.2.

5. If the realization x_2 were not included in Table A.2 (such as $x_2 =$ 0.35% Cu), it would have been necessary to interpolate between available values of x_2 (such as 0.30 and 0.40% Cu).

TABLE A.4 Cumulative probability distribution of x_2 given x_1 = 0.30% Cu

Gold Grade	x_2 (g/t Au)	0.01	0.02	0.04	0.08	0.16
Probability of occurrence	$p(x_2 \| x_1 = 0.30)$	4%	20%	39%	28%	9%
Cumulative probability	$cpd(x_2 \| x_1 = 0.30)$	4%	24%	63%	91%	100%

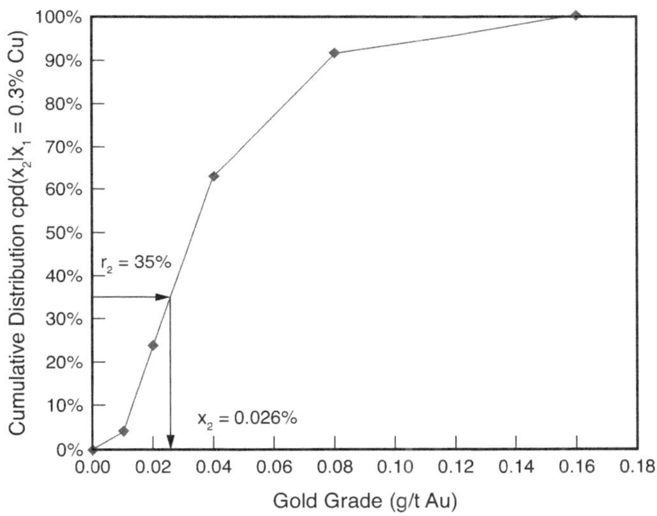

FIGURE A.2 Cumulative probability distribution of x_2 given x_1 = 0.3% Cu

6. Generate a random number between 0 and 1 from a uniform distribution. For this example, it is assumed that the generated random number is r_2 = 35%.

7. Determine the value of x_2 for which $cpd(x_2|x_1 = 0.31\%\ Cu) = r_2$. As shown in Figure A.2, this value is x_2 = 0.026 g/t Au.

8. The values x_1 = 0.31% Cu and x_2 = 0.026 g/t Au represent a single realization of what the deposit grades might be. Using these values (and similarly simulated values of other input variables such as tonnage, metallurgical recovery, etc.), one can calculate the corresponding project utility $U(x_1,x_2)$.

9. The process should be repeated 100 times, using different randomly generated values of r_1 and r_2, thus obtaining 100 realizations of x_1, x_2

and $U(x_1,x_2)$. The probability distribution of $U(x_1,x_2)$ is calculated from these 100 realizations and used to determine what should be the path forward: Continue exploration or accept the current risk and go ahead with developing a new mine.

As should be evident from this example, the main challenge in applying Monte Carlo simulation to correlated variables is in the determination of correlation tables such as Table A.1. For this reason, commercially available software make it possible to use simplifying assumptions, the correlation table being replaced by a single number, a correlation coefficient.

APPENDIX B

GEOSTATISTICAL SIMULATION OF GOLD PRICES

Relatively simple geostatistical methods can be used to simulate realizations of future gold prices when patterns observed one year are used to predict prices the following year. The underlying assumption is that past behavior is indicative of future behavior, an assumption that is rarely true. The simulation method presented here is based on a statistical analysis of historical gold prices using the variogram, followed by Monte Carlo simulation of future gold prices. This method cannot be used to predict the timing or duration of changes in price directions, such as those that are typical of economic cycles.

STEP 1: GEOSTATISTICAL ANALYSIS OF HISTORICAL GOLD PRICES

The variogram is estimated from past information and consists of comparing the gold price x(u) at time u with the gold price x(u + t) at time u + t. The variogram formula is as follows:

$$\gamma(t) = \Sigma[\,x(u) - x\,(u + t)]^2 \,/\, [2\,n]$$

The sum applies to all pairs of gold values that are separated by the same time lag t, and n is the number of such pairs.

In this example, the gold price variogram was calculated using 2014 daily gold prices. On day u, x(u) is the gold price and x(u + t) is the gold price on day u + t. The results were used to simulate realizations of the 2015 daily gold prices.

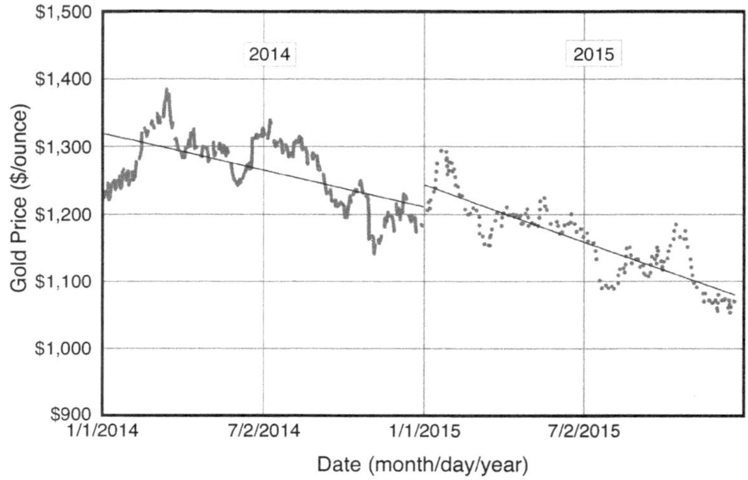

FIGURE B.1 Actual gold prices from January 2014 to December 2015

Figure B.1 shows the daily gold prices for the years 2014 and 2015. Also shown are linear trends for each year. It is evident that one could have used the trend observed in 2014 to reasonably predict prices in 2015. Unfortunately, one never knows with certainty whether trends will continue over time: They usually do not. For the purpose of this study, the assumption is that the 2014 trend could be ignored or was not expected to continue in 2015.

If the 2014 trend had been assumed to continue in 2015, the trend would have been removed from the 2014 prices and the 2014 variogram would have been calculated from the residuals. The 2015 price simulations would have been completed by assuming a known trend (a continuation of that observed in 2014) on which would have been superimposed the simulated residual variations.

The 2014 gold price variogram was calculated for time lags of 1 to 30 days. The results are shown in Figure B.2. As one would expect, the mean squared difference between gold prices is smallest for lags of one day and increases when the time difference between gold prices increases. What might not have been expected is that the variogram increases linearly when the time lag grows. The relationship is represented by the following equation, where t is the time lag measured in days:

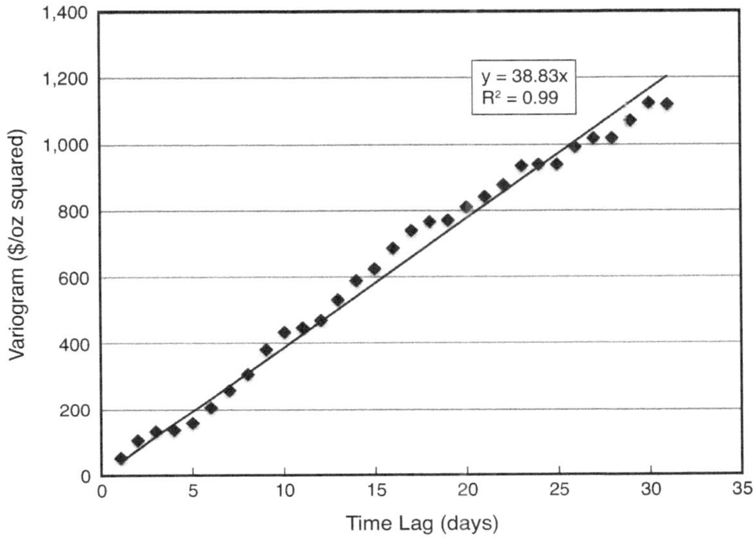

FIGURE B.2 Variogram of daily gold prices in 2014

$$\gamma(t) = 38.83\ t$$

Linear variograms of the form $\gamma(t) = s \times t$, where s is the slope of the fitted model (38.83 in the preceding example), are indicative of a "random walk."[1] The gold price at time u + 1 can be estimated from the gold price at time u using the following equation:

$$x(u + 1) = x(u) + \varepsilon$$

In this equation, ε is a random variable independent of u, with mean $\mu = 0.0$ and variance σ^2. The gold price at time u + t can be written as follows:

$$x(u + t) = x(u) + \Sigma\varepsilon_i,\ i = 1 \ldots t$$

The corresponding variogram is

$$\gamma(t) = t\,\sigma^2 / 2$$

1 Random walks are not uncommon. They have been observed in metal prices as well as in stock market variations.

The variance σ^2 of the random variable ε is equal to twice the slope s of the fitted variogram model: $\sigma^2 = 2$ s.

The standard deviation σ of ε is measured in dollars per ounce. Its value for 2014 is calculated as follows:

$$\sigma = \sqrt{2\ s} = \$8.81 \text{ per ounce}$$

If we assume that ε is approximately normally distributed, there is approximately one chance in three that the gold price will vary by more than \$9 per ounce from one day to the next, and one chance in 20 that the difference will exceed \$18 per ounce. If we know the gold price on a given day, the next-day price is just as likely to be higher as it is to be lower.

A geostatistician might be surprised that the variogram does not have a parabolic shape, as would be expected from the downward trend in gold prices shown in Figure B.1. A simple explanation is that the time lags that were considered in calculating the variogram are less than one month, a period during which prices are sometimes increasing and at other times decreasing.

STEP 2: SIMULATION OF FUTURE GOLD PRICES

The following steps were used to simulate future gold prices using a random walk model:

1. Assume that the daily variations in gold prices, ε, are normally distributed with mean 0.0 and standard deviation \$8.81 per ounce:

$$\text{pd}(\varepsilon) = \text{norm}(0.0{,}8.81)$$

2. The cumulative distribution of ε is cpd(ε):

$$\text{cpd}(\varepsilon) = \text{cum.norm}(0.0{,}8.81)$$

3. Start with the gold price on January 2, 2015:

$$x_0 = \$1{,}189 \text{ per ounce}$$

4. Generate a random number r with uniform distribution between 0 and 1. For this example, assume that r = 33.66%.

5. Calculate the value e of ε whose cumulative probability distribution is equal to r:

$$cpd(e) = 33.66\%$$

   Using the inverse of the cumulative normal distribution cpd(ε), the following value is obtained:

$$e = -\$3.72 \text{ per ounce}$$

6. The simulated realization of the gold price on January 3, 2015, is

$$x_1 = \$1{,}189 - \$3.72 = \$1{,}185 \text{ per ounce}$$

7. Repeat steps 3 to 6, starting with the simulated gold price for January 3, 2015, using a different random number r to simulate the January 4, 2015, gold price.
8. The process is repeated for 365 days. The result is one realization of the possible 2015 gold price.
9. To simulate another realization of the 2015 daily gold prices, it is sufficient to restart with the January 2, 2015, gold price using new random numbers.

Two simulated realizations of the 2015 gold prices are shown in Figure B.3. The starting point is the gold price that prevailed on January 2, 2015. The simulated gold prices are compared with actual prices for the same period. The results show short-term variability similar to that of the actual data. While the day-to-day changes are random, simulations create apparent trends over months, a result which one might not have expected but which is statistically not surprising.

Twelve simulations were completed, which are plotted on Figure B.4. As one would expect, the range of simulated gold prices increases when projections are made later in the year. Although the variogram model was developed for time lags of only 30 days, application of the random walk to simulate future gold prices gives reasonable results over one year, which confirms the stationarity of the price variability during this period. Actual 2015 gold prices fall within the range of simulated realizations. The simulations predicted that the end-of-2015 gold price could differ from that in January by $200/ounce, either up or down. The actual December gold price was $129/ounce lower.

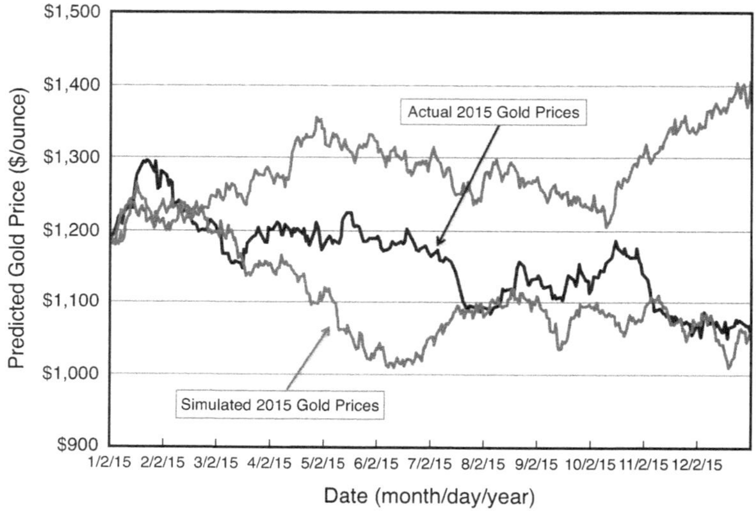

FIGURE B.3 Comparison of actual 2015 gold prices with two selected simulations

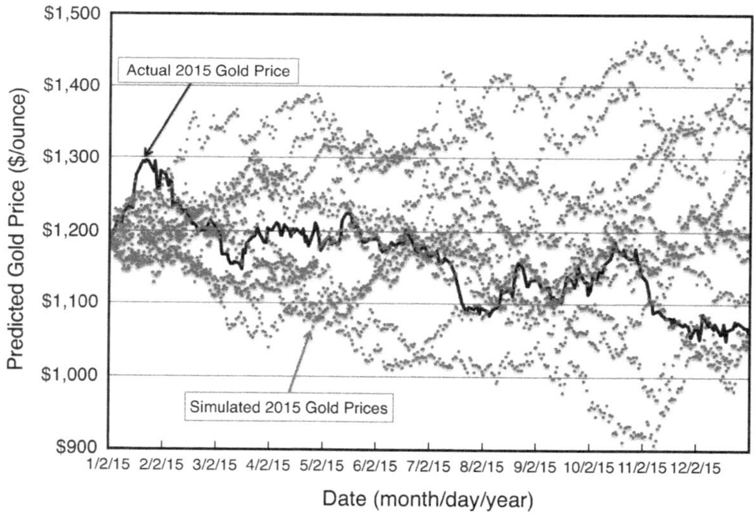

FIGURE B.4 Comparison of actual 2015 gold prices with 12 simulated gold prices

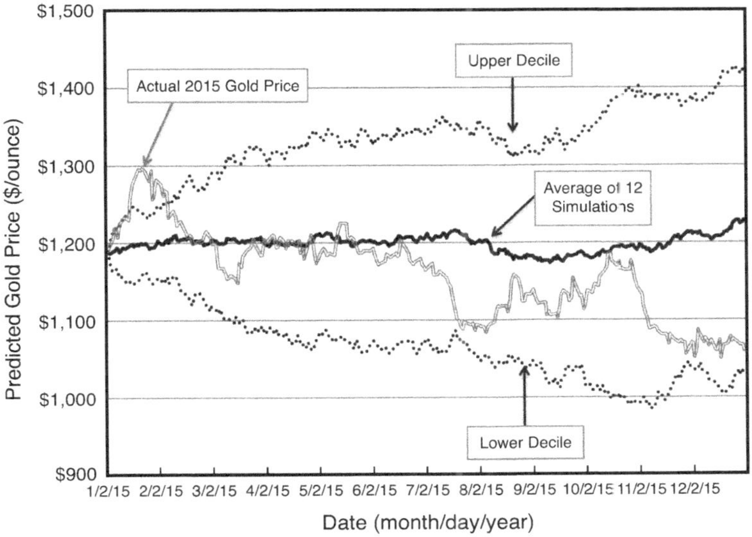

FIGURE B.5 Comparison of actual 2015 gold prices with simulated prices using 2014 statistics

For each day, each one of the 12 simulated realizations predicted a different gold price. The average of these 12 values was calculated, which theoretically, for an infinite number of simulations, should be equal to the initial gold price, that of January 2, 2015. Also calculated was the standard deviation of the 12 values. Using the daily average and standard deviation, confidence intervals were calculated for the projected daily gold prices. A t-distribution with 11 degrees of freedom was assumed. For example, on June 2, 2015, the 12 simulated gold prices averaged $1,200/ounce with a standard deviation of $93/ounce. The calculated confidence intervals for the lower and upper 10th percentiles are $1,073 and $1,326 per ounce. There is 80% probability that the actual gold price will be between these two values.

The daily averages, upper and lower deciles and actual gold prices, are plotted on Figure B.5. As expected, the average of the 12 simulations remains close to the initial gold price of January 2, 2015 at $1,189 per ounce. The confidence limits become larger when projections are made later in the year. If this model had been used in January 2015 to predict December 2015 gold prices, one would have estimated a 10% probability that prices would be less than $1,000 and an equal 10% probability that they would be greater than $1,400 per ounce.

With the exception of January 2015, the actual gold prices fall well within the upper and lower limits. The downward price trend observed in 2014, which continued in 2015, was not taken into account when simulating the 2015 gold prices. If one had assumed that the trend observed in 2014 could be projected forward, the 2015 predictions would have been more accurate. But would it have been reasonable to assume in January 2015 that the 2014 trend would continue?

It is worthwhile noting that although the gold price was $1,060 at the end of 2015, it exceeded $1,300 per ounce by the end of August 2016 (Figure 8.2). While this value is within the ranges predicted in Figure B.5, the sharp change in direction, from slowly decreasing prices to rapidly increasing, would not have been predicted using the above random-walk simulation method. How to simulate such "disruptive" changes was discussed in Chapter 8.

REFERENCES

CCG (Centre for Computational Geostatistics). 2014. Home page. University of Alberta, Canada. www.ccgalberta.com.

COSMO Stochastic Mine Planning Laboratory. n.d. Home page. McGill University, Canada. http://cosmo.mcgill.ca (accessed Nov 2016).

CRIRSCO (Committee for Mineral Reserves International Reporting Standards). 2013. *CRIRSCO International Reporting Template for the Reporting of Exploration Results, Mineral Resources and Mineral Reserves*. www.crirsco.com/template.asp.

DMDU (Society for Decision Making Under Deep Uncertainty). n.d. Home page. www.deepuncertainty.org (accessed Nov. 2016).

The Economist. 2009. Triple bottom line. Nov. 17 (online extra). www.economist.com/node/14301663.

FASB (Financial Accounting Standards Board). 2008. EITF abstracts. Mining Assets: Impairment and Business Combinations. Issue No. 04-3. www.fasb.org/jsp/FASB/Document_C/DocumentPage?cid=1218220141031&acceptedDisclaimer=true.

Graff, R.P. 2008. Recognition of value beyond proven and probable reserves in business combinations. *Min. Eng.* 60(8):45.

ICCPM (International Centre for Complex Project Management). 2014. What is complex project management? *Connect* (newsletter), September. https://iccpm.com/content/september-2014.

ICMM (International Council on Mining and Metals). 2015. www.icmm.com/en-gb/about-us/member-commitments/icmm-10-principles

IRM (Institute of Risk Management). 2016. Definition of *risk*. www.theirm.org/about/risk-management/.

ISO 31000:2009. *Risk Management: Principles and Guidelines*. www.iso.org/iso/home/standards/iso31000.htm.

Journel, A.G., and Huijbregts, Ch.J. 1978. *Mining Geostatistics*. London; New York: Academic Press.

Leuangthong, O., and Deutsch, V. 2003. Stepwise conditional transformation for simulation of multiple variables. *Math. Geol.* 35(2).

NI 43-101. *National Instrument 43-101: Standards of Disclosure for Mineral Projects*. Ontario Securities Commission. www.bcsc.bc.ca/For_Companies/Mining/

Rand Corporation. 2016. Decision Making Under Uncertainty Research Team. www.rand.org/jie/infrastructure-resilience-environment/projects/improving-decisions/researchteam.html.

Rendu, J.M. 2014. *An Introduction to Cut-off Grade Estimation*, 2nd ed. Englewood, CO: SME.

Sarbanes–Oxley Act. 2002. Public Law 107-204, 116 Stat. 745. Section 404, Management assessment of internal controls. https://www.sec.gov/about/laws/soa2002.pdf.

SRA (Society for Risk Analysis). 2016. Home page. www.sra.org (accessed Nov. 2016).

UN (United Nations). 1987. *Report of the World Commission on Environment and Development: Our Common Future (the Brundtland Commission).* Oxford: Oxford University Press.

ABOUT THE AUTHOR

Jean-Michel (J-M) Rendu is an independent consultant supplying services to the international mining industry. J-M retired from Newmont Mining Corporation in 2001 as vice president of resources and mine planning with worldwide responsibility for corporate and mine-site mining engineering activities, and for estimation and reporting of resources and reserves. Before joining Newmont, J-M was an associate consultant with Golder Associates; an assistant professor with the University of Wisconsin–Madison; head of operations research with Anglovaal, Johannesburg, South Africa; and systems analyst with Kennecott Copper Corporation.

J-M graduated from École des Mines de Saint-Étienne in 1966 and obtained M.S. and Eng.Sc.D. degrees from Columbia University in the City of New York. He authored more than 50 technical papers on deposit modeling, mine engineering, estimation of resources and reserves, and U.S. and international regulatory requirements for public reporting. He is the author of *An Introduction to Geostatistical Methods of Mineral Evaluation*, published by the South African Institute of Mining and Metallurgy in 1978 (second edition, 1981) and *An Introduction to Cut-off Grade Estimation*, published by SME in 2008 (second edition, 2014).

In addition to being a founding registered member of SME, J-M has chaired SME's Ethics Committee, Resources and Reserves Committee, and Mining and Exploration Division. He was a founding member and U.S. representative of the Committee for Mineral Reserves International Reporting Standards (CRIRSCO) and is a fellow of the Australasian Institute of Mining and Metallurgy and the Southern African Institute of Mining and Metallurgy.

J-M was a recipient of the Henry Krumb Lecturer Award, the SME Presidential Award, the AIME Mineral Economics Award, the SME Distinguished Member Award, and the American Mining Hall of Fame Medal of Merit. J-M is an elected member of the U.S. National Academy of Engineering.

INDEX

NOTE: *f.* indicates figure; *t.* indicates table